本书由广州市产学研协同创新重大专项：健康产业跨境电子商务平台建设及产学研一体化研究（编号：201704030006），广东省一流高职院校高水平专业（健康管理）建设项目（编号：501C041928G），广东省教育厅课题：多方协同构建中药文化涉外培训体系（编号：20130201045），校级课题（编号：2017ZR009）及广州市岭南中草药科普基地提供资助支持。

岭南中草药资源概览

OVERVIEW OF LINGNAN CHINESE HERBAL MEDICINE RESOURCES

主编：宋　卉　黄珊珊　王少静
参编：沈小钟　郑婷菲　陈拥珊　蔡树坚　蔡岳文
主审：郑彦云　王玉生

广东高等教育出版社

Guangdong Higher Education Press

·广州·

图书在版编目（CIP）数据

岭南中草药资源概览 / 宋卉，黄珊珊，王少静主编. —广州：广东高等教育出版社，2019.3

ISBN 978 – 7 – 5361 – 6236 – 5

Ⅰ. ①岭… Ⅱ. ①宋… ②黄… ③王… Ⅲ. ①药用植物 – 植物资源 – 介绍 – 广东 Ⅳ. ① S567.019.265

中国版本图书馆 CIP 数据核字（2018）第 188373 号

岭南中草药资源概览
Lingnan Zhongcaoyao Ziyuan Gailan

出版发行	广东高等教育出版社
	地址：广州市天河区林和西横路
	邮编：510500　　营销电话：（020）87553735
	网址：www.gdgjs.com.cn
印　　刷	广东信源彩色印务有限公司
开　　本	787 mm × 1 092 mm　　1/16
印　　张	9.5
字　　数	175 千
版　　次	2019 年 3 月第 1 版
印　　次	2019 年 3 月第 1 次印刷
定　　价	36.00 元

前言

自神农氏尝百草到如今21世纪，中草药已经陪伴中华民族走过了5 000年的风风雨雨。中草药文化根植于中国传统文化的土壤之中，融汇了中国5 000年哲学、文学、历史、天文和地理等多学科知识。岭南中草药文化作为中草药文化的一个分支，其发展有得天独厚的地理位置和广泛的人文基础的支持。

岭南地貌类型复杂多样，山地、丘陵、台地、平原交错，海岸线曲折，岛屿众多。位于东亚季风气候区南部，具有热带、亚热带季风海洋性气候特点，高温多雨为其主要气候特征。太阳辐射量较多，日照时间较长，以广东省为例，全省各地平均年日照时数为1 450~2 300小时。因全年气温较高，加上雨水充沛，所以林木茂盛、四季常青，百花争艳，各种果实终年不绝。

岭南特有的地理环境与气候孕育了丰富的中草药资源。据统计，岭南地区药用资源超过4 500种，占全国药用资源种类的36%，其中陆地药用资源中植物类约有4 000种。广东省有2 654种，广西壮族自治区有4 637种，海南省有497种，其中还包括许多特有种类。岭南人民在长期的生产劳动过程中积累了大量的有关药用动植物的采集、分类、栽培、养殖和防病治病等方面的丰富经验，经过去伪存真、优胜劣汰、择优而立的历史性筛选，巴戟天、何首乌、广陈皮、广佛手、化州橘红、广东金钱草、广藿香、肉桂、槟榔、砂仁、益智仁、高良姜等品质优良的传统中药，成为公认的"岭南道地药材"。

认识和鉴别中草药资源是掌握中草药知识的第一步，相关教材与科普读物是最好的媒介与桥梁。目前，国内有关中草药资源的教材品类众多，但是能够兼顾中英文的双语教材十分稀少，相关的科普读物亦是如此，这与国际社会对中医药日益高涨的学习需求相矛盾。因此，本书既能让国内青少年学生掌握中草药知识以及中英文双语的规范表达，为其以后参与国际交流奠定基础；也可进一步扩大中草药药材在国际上的影响。

《岭南中草药资源概览》是一本颇具特色的岭南中草药资源概述性图书，以中英文双语的表达形式，附加彩色图片资料，对岭南常见中草药资源进行概述，使初学者对岭南中草药有直观的体验与了解，也兼顾海外读者学习需要。

本书收录岭南常见中草药资源55种，按科属、药名、别名、植物特征、资源状况、药用部位、功能主治、主要成分等内容进行描述。本书所记述的中草药都经过文献查证，所属的种、属均由科学鉴定所确定。性状描述根据其形态特征，以精练的文字表达。功能主治和主要成分源于中医药文献的报道。对应的英文翻译由广东食品药品职业学院长期聘请的外教Todd Smith（托德·史密斯）审定，保证了英文用语的正确性与严谨性。

本书中文部分由黄珊珊、沈小钟编写，郑彦云主任药师、王玉生主任中药师负责主审；英文部分由王少静、郑婷菲、陈拥珊等翻译，宋卉教授、Todd Smith统一审核定稿，药用植物图片主要由蔡岳文主任中药师和蔡树坚提供。

本书以科普为主要目的，内容比较简要，希望广大读者能有所收获。

编　者

2018年10月

Preface

Since the Agriculture Emperor's Materia Medica first tasted the herbs to the 21st century, Chinese herbal medicine has been accompanying with the Chinese history for more than five thousand years. The culture of Chinese herbal medicine is rooted in the traditional Chinese culture, and has been integrated with many disciplines like Chinese philosophy, literature, history, astronomy and geography. As a branch of Chinese herbal medicine culture, Lingnan Chinese herb culture development has been supported by its unique geographical position and extensive humanistic foundation.

The types of Lingnan landform are complex and various, which intersect with mountains, hills, terraces and plains, zigzags coast and numerous islands. It belongs to south east Asian monsoon climate zone, with tropical and subtropical monsoon oceanic climate characteristics. High temperature and rainfall are the main climatic features. It has a fairly solar radiation quantity and long sunshine time. Taking Guangdong province as an example, the average sunshine time of the whole province is between 1,450 h to 2,300 h. Due to the high temperature of the year and the abundant rain, it has flourishing trees, four evergreen seasons, blooming flowers and plentiful fruits all year round.

The unique geographical environment and climate of Lingnan make it rich in traditional Chinese herbal medicine resources. According to statistics, there are more than 4,500 medicinal resources in Lingnan area, which accounts for 36% of the national medicinal resources. Among them, there are about 4,000 kinds of medicinal plants are land resources. There are 2, 654 species in Guangdong province, 4,637 species in Guangxi Zhuang Autonomous Region and 497 species in Hainan province, which include many unique species. The Lingnan people have accumulated a great deal of experience in the collection, classification, cultivation, breeding and disease prevention and treatment of medicinal plants and animals in the long process of production and labor. After the historical screening of the fittest, the survival of the fittest, the selection of the best, some high quality traditional Chinese medicines like *Morinda officinalis* How, *Fallopia multiflora* (Thunb.) Haraldson, *Citrus reticulata* Blanco, *Citrus medica* L.var. Sarcodactylis Swingle, *Citrus maxima*' Tomentosa', *Desmodium styracifolium* (Osbeck) Merr., *Pogostemon cablin* (Blanco) Benth., *Cinnamomum cassia* Presl, *Areca catechu* Linn., *Amomum villosum* Lour., *Alpinia oxyphylla* Miq. and *Alpinia officinarum* Hance are widely recognized as "the medicinal herb of Lingnan".

Understanding and identifying Chinese herbal medicine resources is the first step to master the knowledge of Chinese herbal medicine. Relevant textbooks and popular science books are the good media and bridge for learning. At present, there are many textbooks about Chinese herbal medicine resources in China, but the bilingual textbooks and the relevant popular science books in both Chinese and English are very limited. This contradicts the international community's growing demands for Chinese medicine. Therefore, this book intends to meet the international demands of preliminary understanding of Lingnan medicinal materials, and more importantly, it intends to make the domestic young students master the specific expressions of the Chinese herbal medicine in Chinese and English. So as to lay foundation for them to communicate with others, to further expand the influence of Chinese traditional medicine in the world.

Overview of Lingnan Chinese Herbal Medicine Resources is an overview book of Lingnan traditional Chinese medicine resources, with a large number of pictures to summarize the common Chinese medicinal materials in Lingnan in the form of Chinese and English. It is used to give beginners an intuitive experience and understanding of Chinese herbal medicine, and also to serve the overseas readers.

This book contains 55 common Lingnan Chinese medicinal herbs, which are described in terms of genus, scientific names, features, resources situation, medicinal parts, functions, main ingredients, etc. All the Chinese herbal medicines described in this book have been verified by literature, and their species and genera are confirmed by Scientific Appraisal Institute. The English translation is mainly proofed by Todd Smith, the English teacher hired by Guangdong Food and Drug Vocational College for a long time to make sure the accuracy and preciseness of the language.

The Chinese draft is written by Huang Shanshan, Shen Xiaozhong, mainly proofed by Professor of Pharmacy Zheng Yanyun, Professor of medicine Wang Yusheng. The English translation is written by Wang Shaojing, Zheng Tingfei and Chen Yongshan. The general review and proof are done by Professor Song Hui, Todd Smith. Pictures of medicinal plants are provided by Professor of Medicine Cai Yuewen and Cai Shujian.

And the main purpose of this book is for popular science reading. The content is simple, and we really hope that the readers can learn the knowledge from it.

editor

October 2018

目　录

第一章　岭南中药资源概述 …………………… 1
　1. 岭南中药资源分布 …………………………… 2
　2. 广东省中药资源区 …………………………… 5
　3. 岭南主要特产药材及生产现状 ……………… 6

第二章　岭南特产中草药资源 ………………… 23
　1. 佛　手 ………………………………………… 24
　2. 广藿香 ………………………………………… 26
　3. 春砂仁 ………………………………………… 28
　4. 巴戟天 ………………………………………… 30
　5. 沉　香 ………………………………………… 32
　6. 高良姜 ………………………………………… 34
　7. 化州橘红 ……………………………………… 36
　8. 何首乌 ………………………………………… 38
　9. 肉　桂 ………………………………………… 40
　10. 广陈皮 ……………………………………… 42
　11. 益智仁 ……………………………………… 44
　12. 降　香 ……………………………………… 46
　13. 槟　榔 ……………………………………… 48

第三章　岭南主产中草药资源 ········· 51

1. 广东金钱草 ········· 52
2. 蔓荆子 ········· 54
3. 穿心莲 ········· 56
4. 溪黄草 ········· 58
5. 龙脷叶 ········· 60
6. 山　奈 ········· 62
7. 蔓性千斤拔 ········· 64
8. 岗梅根 ········· 66
9. 凉粉草 ········· 68
10. 牛大力 ········· 70
11. 香　茅 ········· 72
12. 五指毛桃 ········· 74
13. 黄皮核 ········· 76

第四章　岭南引种进口中草药资源 ········· 79

1. 檀　香 ········· 80
2. 儿　茶 ········· 82
3. 大风子 ········· 84
4. 马钱子 ········· 86
5. 诃　子 ········· 88
6. 苏　木 ········· 90
7. 胖大海 ········· 92

第五章　岭南其他药用植物 ········· 95

1. 见血封喉 ········· 96
2. 喜　树 ········· 98

3. 大叶骨碎补 …………………………………… 100

4. 狗　脊 …………………………………………… 102

5. 贯　众 …………………………………………… 104

6. 威灵仙 …………………………………………… 106

7. 阴　香 …………………………………………… 108

8. 土茯苓 …………………………………………… 110

9. 射　干 …………………………………………… 112

10. 郁　金 ………………………………………… 114

11. 莪　术 ………………………………………… 116

12. 余甘子 ………………………………………… 118

13. 断肠草 ………………………………………… 120

14. 土牛膝 ………………………………………… 122

15. 穿破石 ………………………………………… 124

16. 九里香 ………………………………………… 126

17. 三叉苦 ………………………………………… 128

18. 紫金牛 ………………………………………… 130

19. 白花蛇舌草 …………………………………… 132

20. 广防己 ………………………………………… 134

21. 防风草 ………………………………………… 136

22. 八角枫 ………………………………………… 138

参考文献 …………………………………………… 140

Contents

Chapter 1 Overview of Lingnan Chinese medicine resources 9

1. The location of Lingnan traditional Chinese medicine resources 12
2. Traditional Chinese medicine resource distribution areas of Guangdong Province 16
3. Lingnan major specialty herbal medicines and their production status 18

Chapter 2 Lingnan specialty Chinese herbal medicine resources 23

1. *Citrus medica* L. var. *sarcodactylis* Swingle 25
2. *Pogostemon cablin* (Blanco) Benth. 27
3. *Amomum villosum* Lour. 29
4. *Morinda officinalis* How 31
5. *Aquilaria sinensis* (Lour.) Spreng. 33
6. *Alpinia officinarum* Hance 35
7. *Citrus maxima*' Tomentosa' 37
8. *Fallopia multiflora* (Thunb.) Haraldson 39
9. *Cinnamomum cassia* Presl 41
10. *Citrus reticulata* Blanco 43
11. *Alpinia oxyphylla* Miq. 45
12. *Dalbergia odorifera* T. Chen 47
13. *Areca catechu* Linn. 49

Chapter 3 Lingnan major Chinese herbal medicine resources 51

1. *Desmodium styracifolium* (Osbeck) Merr. 53
2. *Vitex trifolia* Linn. 55
3. *Andrographis paniculata* (Burm.f.) Nees 57
4. *Isodon serra* (Maxim.) Kudo 59
5. *Sauropus spatulifolius* Beille 61
6. *Kaempferia galanga* Linn. 63
7. *Flemingia philippinensis* Merr.et Rolfe 65
8. *Ilex asprella* (Hook. et Arn.) Champ. ex Beth. 67
9. *Mesona chinensis* Benth. 69
10. *Millettia speciosa* Champ. 71
11. *Cymbopogon citratus* (DC.) Stapf 73
12. *Ficus hirta* Vahl 75
13. *Clausena lansium* (Lour.) Skeels 77

Chapter 4 Lingnan import Chinese herbal medicine resources 79

1. *Santalum album* Linn. 81
2. *Acacia catechu* (L.f.) Willd. 83
3. *Hydnocarpus anthelminticus* Pierre ex Lanessan 85
4. *Strychnos nux-vomica* Linn. 87
5. *Terminalia chebula* Retz. 89
6. *Caesalpinia sappan* Linn. 91
7. *Sterculia lychnophora* Hance 93

Chapter 5 Lingnan other medicinal plants 95

1. *Antiaris toxicaria* Lesch. 97
2. *Camptotheca acuminata* Decne. 99

3. *Davallia formosana* Hayata ……………………………………………… 101

4. *Cibotium barometz* (Linn.) J.Sm. ……………………………………… 103

5. *Cyrtomium fortunei* J. Sm. ……………………………………………… 105

6. *Clematis chinensis* Osbeck …………………………………………… 107

7. *Cinnamomum burmanni* (Nees et T.Nees) Blume …………………… 109

8. *Smilax glabra* Roxb. …………………………………………………… 111

9. *Belamcanda chinensis* (L.) Redouté. ………………………………… 113

10. *Curcuma aromatica* Salisb. ………………………………………… 115

11. *Curcuma zedoaria* (Christm.) Rosc. ………………………………… 117

12. *Phyllanthus emblica* Linn. …………………………………………… 119

13. *Gelsemium elegans* (Gardn.et Champ.) Benth. …………………… 121

14. *Achyranthes aspera* Linn. …………………………………………… 123

15. *Maclura tricuspidata* Carrière ……………………………………… 125

16. *Murraya exotica* L. …………………………………………………… 127

17. *Evodia lepta* (Spreng.) Merr. ………………………………………… 129

18. *Ardisia japonica* (Thunb.) Blume …………………………………… 131

19. *Hedyotis diffusa* Willd. ……………………………………………… 133

20. *Aristolochia fangchi* Y.C.Wu ex L.D.Chow et S.M.Hwang ………… 135

21. *Epimeredi indica* (L.) Rothm. ………………………………………… 137

22. *Alangium chinense* (Lour.) Harms …………………………………… 139

第一章
岭南中药资源概述

岭南中草药发展历史悠久。自晋代以来，以葛洪为代表的医药名家辈出，如支法存、鲍姑、仰道人等推动了岭南医学的迅速发展，也极大地提高了社会对中草药的需求与探索。其间各类岭南中草药相关著作的问世，彰显出岭南中草药的地位与重要性。如《南裔异物志》为汉代杨孚以古诗写成的首部具有岭南独特风格的风物志，是岭南地区最早的动物、植物志。该书精确描述了岭南特有的动物、植物特性，对岭南药物学的发展具有一定的影响力。《海药本草》为五代李珣所著，是介绍以海外药物为主的本草著作，收录了海外产地药96种，如青木香、荜拨、红豆蔻、丁香、乳头香、没药、甘松香等。此外，还有明清时期王纶的《本草集药》、赵寅谷的《本草求原》、肖步丹的《岭南采药录》、胡真的《山草药指南》等。

岭南地处五岭（大庾岭、骑田岭、萌渚岭、都庞岭、越城岭）以南，逶迤分布于湘、赣、粤、桂之间，其所辖范围约为我国的广东、香港、澳门、海南及广西大部和越南北部。岭南北倚五岭，南临大海，属热带、亚热带气候，雨量充沛，地形复杂，地貌多样，海洋陆地兼备，适合各种动植物生长，因此，中药资源不仅品种多、分布广、产量大，而且有不少特产品种质量好，驰名国内外。

岭南家种药材产量和野生药材蕴藏量均十分丰富。据有关资料统计，蕴藏量在100万千克以上的大宗品种药材有近300种。历史上所形成的地道药材，如德庆首乌、连州玉竹、西江桂皮、河源五指毛桃、新会陈皮、石牌藿香、化州橘红、砂仁、广佛手、巴戟天、沉香、高良姜、金花茶、广西大青、广地龙、白花蛇等品种在中华人民共和国成立前已远近驰名，行销国内外。此外，肉桂、檀香、银杏、杜仲、黄檗、厚朴、吴茱萸、蔓荆子、山药、泽泻、天花粉、使君子、穿心莲、紫苏、广东金钱草、鸡骨草、薄荷、水半夏、干姜、姜黄、山柰、壳砂仁、黄精、龙脷叶、芦荟、茯苓、灵芝等是常见的家种药材品种。岭南野生药用植物主要有沉香、淡竹叶、山银花、土茯苓、木姜子、海金沙、鸦胆子、白茅根、山芝麻、广防风、毛冬青、金樱子、鱼腥草、香附子、鬼针草、灯芯草、南五味子等。野生及饲养的药用动物有蛤蚧、水蛭、地龙、蝎子、金沙牛、蚝、蚕等，还有大量的海洋药用生物，如海马、海龙、海星、海藻、鹧鸪菜及贝壳类动物。

岭南中药素有"广药"之称。据考证，岭南地区药用资源超过4 500种，占全国药用资源种类的36%，其中广东省药用资源种类占全国的20.7%，而广西壮族自治区药材资源品种总数位列全国第二，仅次于云南省。根据20世纪80年代全国中药资源普查资料，广东省中药资源有2 614种，其中药用植物2 500种，隶属225科、1 175属；药用动物109种，隶属89科；药用矿物25种。当时编入《广东中药资源名录》的药用植物有1 170种、药用动物109种、药用矿物24种，共1 303种。广西壮族自治区中药资源共有4 637种（包括亚种、变种、变型），隶属548科，1 861属，其中藻类15种12科12属；真菌类85种28科49属；地衣类10种5科7属；苔藓类15种12科13属；蕨类225种46科89属；药用植物3 728种221科1 343属；药用动物509种214科348属；药用矿物50种。丰富的中药品种资源，为当地中药制药产业的形成和发展提供了强大的物质基础条件，仅广东省具有一百年以上历史的老字号知名中药制药企业就有近10家，如陈李济、冯了性、白云山中一药业、敬修堂、王老吉及潘高寿等。

1. 岭南中药资源分布

（1）药用植物

植物生长与生态环境有着密切的关系。岭南地区虽然没有严寒的冬季，但气候、土壤和植被都呈明显的带状分布。因此，中药资源分布也有带状分布的特征。现将岭南主要植物药材品种，按其产地的地理位置和自然条件分述如下。

① 热带地区

广东省西南部、广西壮族自治区东南部及海南省属热带地区，年积温8 000~8 500 ℃，长夏无冬。年降雨量1 500~2 500 mm，具热带季风气候的特点。地带性土壤为砖红壤。植被类型为热带雨林、热带季雨林、红树林、热带草原和热带海滨沙生植物等。

从垂直分布上看，由于地形地貌以及植被类型的差异，植物药材品种分布亦有所不同。广东雷州半岛海拔在100 m以下，地表起伏和缓，是近代玄武岩台地。本地区主要有低丘台地、海滨沙滩和红树林带。分布的主要药用植物资源有龙血树、肉桂、芦荟、益智、槟榔、小叶地不容（金不换）、千余藤等。

② 低丘台地

此带人口密集，由于长期的开发利用，原始植被已不存在，大部分是人工栽培植被。分布的植物药材资源主要有：了哥王、大青、山芝麻、天冬、千斤拔、牛大力、鸡骨香、布渣叶、桃金娘、救必应、葫芦茶、广东金钱草、土茯苓、海金沙、鸦胆子、土牛漆、马鞭草、白茅根、高良姜、海刀豆、穿破石、佛手、香薷、独脚金、土丁桂、毛麝香、香附子、崩大碗、地胆头、鬼针草等野生品种。栽培品种有广藿香、高良姜、山奈、穿心莲、鸡骨草等。

③ 海滨沙滩

此带分布于沿海。大多海拔低于10 m，环境干热，太阳辐射强烈，土壤受海风海水的影响，盐分高。主要植物药材品种有单叶蔓荆、香附子、厚藤、海刀豆、芦荟、长春花、仙人掌、天门冬等。

④ 红树林带

红树林带是一种热带海滩的特殊生态类型，植被为红树林和半红树林群落，植物资源主要有许树、木榄、角果木、海榄齿、海芒果、露兜、草海桐等。

⑤ 南亚热带地区

此带位于我国亚热带的东部，广东的怀集、清远、佛冈、龙川、大埔等市县以南地区，东南临海，西南接雷州半岛北部的热带植被带。境内地形以孤山丘陵为主，其次为三角洲冲积平原。

北回归线横贯本带北部，属于热带季风类型，有明显的干湿季之分。年平均气温为20～22 ℃，极端气温在大寒潮时可降至0 ℃，有轻霜。年降雨量1 500～2 200 mm。土壤为砖红壤性红壤、山地红壤及山地黄壤。植被类型为亚热带常绿季雨林、次生亚热带草坡及人工林。

植被的种类成分具有热带、亚热带的过度类型，但热带植物区系成分占多数。人工栽培品种和野生中药资源均比较丰富。主要有：砂仁、何首乌、化州橘红、白椿、茯苓、泽泻、广藿香、广佛手、紫苏、干姜、郁金、山奈、金银花、射干、千年健、芡实、穿心莲等家种，以及木鳖子、蔓荆子、草豆蔻、巴豆、相思子、葛根、紫花地丁、广防己、伸筋草、鸡骨香、十大功劳、广东金钱草、鸭脚木、夏枯草等野生品种。

⑥ 中亚热带地区

广东的清远、韶关、河源、梅州的北部地区市县以丘陵山地为主，气候年变化幅度较大。年平均气温18～20 ℃，极端气温最低在0 ℃以下，霜期长达一个半月左右，有冰冻，有些年份有下雪现象。年降雨量1 500 mm左右，部分地区可达2 000 mm。春雨早，降雨季节分布均匀，气候较湿润，旱季较短，是广东省野生中药资源最丰富的地区。

从植被的水分分布来看，本地区植被以亚热带区系成分为主，其次为热带区系和山地的种类成分，温带区系成分也不少。主要品种有厚朴、杜仲、三七、玉竹、桔梗、白术、白芷等家种品种，以及马兜铃、远志、翻白叶、女贞子、广东升麻、南丹参、黄精等野生品种。

垂直分布上也有很大的差异，山地主要品种有：黄连、三七、三尖杉、龙胆叶、藁本、天南星、黄精、金耳环、钩藤等。疏林沟边主要品种有：马蓝、鸡血藤、贯众、虎杖、茜草、溪黄草、木通、瓜蒌等。丘陵主要品种有：仙茅、山苍子、乌药、了哥王、毛冬青、岗梅、山栀子、金银花、金樱子、南五味子、黑老虎、野葛、土茯苓等。草坡上主要品种有：苍耳子、山芝麻、紫花地丁、独脚金、鬼羽箭、广东土牛膝等。平原及田野主要品种有：木槿、樟树、桑、苦楝、木芙蓉、佛手、菊花、薄荷、益母草、艾叶、旱莲草、田基黄、车前草、白花蛇舌草、鱼腥草、夏枯草、半边莲等。江河、湖泊及沼泽地主要品种有：莲、芦根、菖蒲等。

（2）药用动物药和药用矿物

① 药用动物药

陆地药用动物有：华南虎、梅花鹿、獐、猴、蛤蚧、石燕、龟、鳖、穿山甲、豹、蛇、地龙、蜂房、蝙蝠、蝉、金沙牛、竹蜂、螳螂、蟾蜍等。海洋生物有：珊瑚、石决明、贝齿、瓦楞子、珍珠贝、海星、海胆、鲨鱼、海马、海龙、海麻雀、海龟、玳瑁、海蛇、海藻、鹧鸪菜等。

② 药用矿物

以广东省为例，主要药用矿物及其分布如表1-1所示。

表1-1 广东省主要药用矿物及其分布

中药名	主要分布市县（区）
浮海石	湛江、珠海、电白、台山、阳江、惠来、潮阳、惠东、海丰、陆丰等
石花蕊	和平
赭石	顺德
石硫黄	英德
石膏	兴宁
寒水石	清远、曲江、始兴、和平、廉江、高州、信宜、蕉岭、梅县等
禹粮石	清远
滑石	连南、高州、信宜、阳春、廉江等
磁石	大埔、阳春、和平、新丰、佛冈等
自然铜	乐昌、阳春、开平、恩平、兴宁等
钟乳石	梅县、蕉岭、阳春、怀集、封开、韶关、惠东、和平、海丰、清远等
紫白石英	清远、河源等
部分矿物药：钟乳石、代赭石、滑石、紫石英、赤石脂	

2. 广东省中药资源区

根据自然地理、资源结构及分布特点，结合中药材生产历史与农业、林业、土壤、气候和植被等专业区划相协调等因素，广东省可分为6个中药资源区。

（1）粤北、粤东北山地、丘陵药材资源区

本药材资源区面积约占广东省全省的1/3。中药资源丰富，南北药材兼备，种类繁多，药用动物资源为全省之冠。药用动物有白花蛇、全蝎、蜈蚣、水蛭、地龙、土鳖虫。种植的主要植物药材有：黄檗、杜仲、厚朴、银杏、玉竹、黄精、茯苓等。野生资源主要有：卷柏、狗脊、贯众、乌药、野葛、金樱子、岗梅、山楂、金银花、溪黄草、淡竹叶等。

（2）粤东南丘陵、台地药材资源区

本药材资源区地貌多样，台地广布。地形北高南低，背山面海。平原和浅海滩涂也占有一定面积，适合各种植物药材生长。栽培的植物药材有穿心莲、广东金钱草、溪黄草、白木香、山柰、高良姜等。该地区也有引种省外药材品种的历史，主要品种有党参、川芎、地黄等。

（3）珠江三角洲药材资源区

本药材资源区最近十多年来经济发展迅速，药用植物种植相对减少。药用植物种植主要有排草、灯芯草、紫苏、素馨花、龙脷叶、红丝线等。野生药用植物资源有紫花杜鹃、金毛狗脊、鸡血藤、骨碎补、白茅根、布渣叶、土茯苓、山芝麻等。

（4）粤西丘陵、山地药材资源区

本药材资源区有多种广东特产药材，是广东省药材的主要生产地。主要品种有广陈皮、砂仁、化州橘红、广佛手、巴戟天、肉桂、广藿香、何首乌、金银花、龙脷叶、芡实、郁金、姜黄、八角、沉香、檀香、益智仁、山药、茯苓、粉葛等。野生药材也十分丰富。

（5）雷州半岛热带药材资源区

本药材资源区为广东省西南的热带半岛。热带植物药材资源丰富，但因干旱缺水，仅主要种植穿心莲、广东金钱草、鸡骨草等。遂溪岭北镇的湛江南药试验场，20世纪60年代开始引种成功多种进口南药，以檀香为主，还有儿茶、大风子、马钱子、安息香、诃子等。

（6）南海海洋药材资源区

广东省大陆海岸线长 3 368.1 km，居全国第一。海产中药资源十分丰富，仅海丰一带就采集有海产药材标本 108 种，主要品种有海马、海龙、玳瑁、牡蛎、海藻、昆布、海带、石莼、江蓠等，海洋养殖的药用品种有海马、珍珠、鲍鱼等。

3. 岭南主要特产药材及生产现状

（1）广藿香

本品为唇形科植物广藿香 *Pogostemon cablin* （Blanco）Benth.，以全草入药。性味辛、微温，归脾、胃、肺经。具有芳香化浊、开胃止呕、发表解暑等

功能。用于治疗湿浊中阻、脘痞呕吐、暑湿倦怠、寒湿闭暑、腹痛吐泻、鼻渊头痛等。广藿香是中成药的原料，广藿香油是医药工业、轻工（香料、香精、香水）的主要原料，是重要的出口商品。

广藿香原产于印度和马来西亚，我国岭南（今广东省）在宋代或更早已有引种，广西、云南、海南也有栽培。过去广州、肇庆是主要产地，而且质量最好，现在主产地是湛江市、阳江市、茂名市，以及遂溪县、吴川市、阳春市最多。电白、化州、徐闻、廉江、四会、高要等市、县也有种植。

（2）佛手

本品为芸香科植物佛手柑 Citrus medica L.var. sarcodactylis Swingle 的果实，佛手性味辛苦酸，温。入肝、胃经。有疏肝理气、和胃止痛的作用。用于甘味气滞、胸胁胀痛、胃脘痞满、食少呕吐等症，还可解酒。

佛手主要分布于高要、四会、德庆、云浮等县、市，近年来粤东的河源市、梅州市、潮州市以及廉江市也有种植。

（3）檀香

本品为檀香科植物檀香 Santalum album Linn.，心材供药用，行气温中，开胃止痛。用于寒凝气滞，胸痛，腹痛，胃疼食少，冠心病，心绞痛。檀香心材芳香，适用于制作檀香工艺雕刻品。檀香油是名贵香精，价格昂贵。

檀香主产于印度、印度尼西亚、马来西亚等地。我国自1962年开始引种栽培，经过近30年的试验研究，推广栽培，已经总结出一套成熟的栽培技术。目前主要有阳西、德庆、廉江、化州、遂溪、电白等市、县，广州及周边地区也有种植。

（4）巴戟天

本品为茜草科植物巴戟天 Morinda officinalis How，性温，味辛甘。入肝、肾经。有补肾、祛风湿的功效。治阳痿、小腹冷痛、月经不调、子宫虚冷、风寒湿痹、腰膝酸痛等症。

巴戟天主产于广东、广西、福建等地，越南也有产。广东省主要产于高要、德庆、郁南、怀集、五华、河源等市、县，以高要、德庆产量最大。

（5）南肉桂

本品来源于樟科肉桂 Cinnamomum cassia Presl 的一个变形，当时据称为越南的清化肉桂，商品名暂定为南肉桂。药用树皮、嫩枝（桂枝）和幼嫩果实（桂子）。桂皮：补元阳，暖脾胃，除积冷，通血脉。桂枝发表解肌，温经通脉。桂子：温中，暖胃。

南肉桂主要分布于广东信宜、高要、罗定、德庆、郁南、云浮等县、市，至1986年年底广东省种植面积140多公顷，信宜种植面积最大，高要、罗定、德庆等，西江流域以西江肉桂为主。

（6）春砂仁

本品为姜科植物春砂仁 *Amomum villosum* Lour. 的果实，是中药砂仁中的佳品。性温、味辛。归脾、胃、肾经。具有化湿开胃、温脾止泻、理气安胎的功效。用于治疗湿浊中阻、脘痞不饥、脾胃虚寒、呕吐泄泻、妊娠恶阻、胎动不安等症。

春砂仁是我国特产中药材，历来以广东阳春出产最为地道，云南、海南也有种植，广东除阳春外还有信宜、高州、广宁、封开等有产。

（7）高良姜

本品为姜科植物高良姜 *Alpinia officinarum* Hance 的干燥根茎。高良姜性热、味辛。归脾、胃经。具有温胃散寒、消食止痛的功效。用于治疗脘腹冷痛、胃寒呕吐、嗳气吞酸等症。

高良姜主产于徐闻、雷州、遂溪、廉江，以徐闻县的龙塘、附城、曲介、前山、锦和等乡镇面积较大，质量也较好。

（8）山柰

本品为姜科山柰属植物山柰 *Kaempferia galanga* Linn. 干燥根茎，又名沙姜，性温、味辛，入胃经，具有温中、消食、止痛的功效。用于胸膈胀满、脘腹冷痛、饮食不消、跌打损伤、牙痛。另外，山柰还大量用于副食品的调味原料。

山柰在广东省分布较广，在粤东的梅州、河源、惠州、潮州地区，粤西的湛江、茂名、阳江、肇庆也有种植。

（9）化州橘红

本品为芸香科植物化州柚 *Citrus granolis*（L.）的果皮，入脾经。具有散寒、燥湿、利气、消痰止咳的功效。用于治疗风寒咳嗽、喉痒痰多、食积伤酒、呕吐恶痞闷等症。国内有20多家药厂以化州橘红为原料生产中成药，化州产的正毛化橘红供不应求，药厂多数以柚皮代替，影响药的质量。

化州橘红主产于化州市，在本地有几百年的种植历史，此外廉江、遂溪、茂名等地亦有产，但以化州产的正毛化橘红质量最佳。

（10）广陈皮

本品来源于芸香科植物茶枝柑 *Citrus chachiensis* Hoet.、蕉柑 *C.tankan* Hayata. 等多种柑橘类成熟果皮。陈皮性温，味辛苦，归肺、脾经，具有理气健脾、燥湿化痰的功效，用于胸脘胀满、食少吐泻、咳嗽痰多。

广陈皮是广东"十大广药"之一，主产地有新会、四会、潮州、博恩、普宁等。广东又以新会陈皮质量最优，是最主要的出口产品。

（11）海马

本品为海龙科动物克氏海马、刺海马、大海马、斑海马或日本海马除去内脏的全体，性温、味甘，归肝、肾经。具有补肾壮阳、调气活血的功效，用于治疗阳痿、遗尿、虚喘、难产、疔疮肿毒等症。

海马分布于热带及温带沿海内湾，风平浪静、水质清、藻类繁衍、浮游动物丰富的低潮区。广东省的陆丰、雷州、电白、吴川等县、市有人工养殖历史。

（12）广地龙

本品为钜蚓科动物参环毛蚓 *Pheretima asperfillm*（Perrier）的干燥全体，习称"广地龙"，性寒、味咸。归肝、脾、膀胱经，具有清热定惊、通络、平喘、利尿的功效，用于治疗高热神昏、惊痫抽搐、关节痹痛、肢体麻木、半身不遂、肺热喘咳、尿少水肿、高血压等症。

广地龙是野生品种，主产于广东佛山、江门、惠州、河源、梅州、茂名等地区。鹤山、南海历来有加工广地龙出口的历史。

Chapter 1 Overview of Lingnan Chinese medicine resources

Lingnan traditional Chinese medicine has a long history. Since Jin dynasty, there have been emerged a great number of famous medical professionals who represented by Ge Hong, and other professionals like Zhi Facun, Bao Gu, and Yang Daoren to promote the rapid development of Lingnan medicine. It also greatly improved the demand and exploration of Chinese herbal medicine. Later, with the introduction of all kinds of related works of Lingnan Chinese herbal medicine, the status and importance of Chinese herbal medicine in Lingnan were highlighted. For example, *Southern Foreign Object Chronicles* was the first narration with Lingnan's unique style which was written with the ancient poem by Yang Fu in Han dynasty. It is the earliest narration of animals and plants in Lingnan area. It accurately describes the characteristics of Lingnan specific animals and plants, and has a certain influence on the development of Lingnan pharmacology. *Overseas Herbal* was written by Li Xun of Five Dynasties. It was an introduction to the overseas medicine, which contained 96 kinds of overseas medicines, such as *Cocculus orbiculatus*, *Piper longum*, *Alpinia galanga*, *Ocimum gratissimum* Var. *suave*, *frankincense*, *Commiphora myrrha*, *Nardostachys jatamansi*, etc. In addition, there are other famous books including *Herbal Medicine* written by Wang Lun in Mingqing dynasty, *The Origin of Herbal* written by Zhao Yangu, *Lingnan Extracts* written by Xiao Budan, *Mountain Herbal Guide* written by Hu Zhen, etc.

Lingnan is located in south of Five Ridges, which include Dayu Ridge, Qitian Ridge, Mengzhu Ridge, Dupang Ridge and Yuecheng Ridge. Five Rideges are in the area of Hunan, Jiangxi, Guangdong and Guangxi Province. Lingnan covers the area of Guangdong, Hong Kong, Macao, Hainan, most part of Guangxi China and the north part of Vietnam. The north part of Lingnan is close to Five Ridges. The south part of Lingnan is next to the sea. Lingnan belongs to tropical and subtropical regions, which has abundant rainfall, complex terrain, and covers the sea area and the land area. This region is suitable for various plants and animals. A variety of traditional Chinese medicine resources has been discovered in wild areas and has an abundant output. There are many varieties of good quality products which are well known at home and abroad.

Lingnan has abundant production and wild medicinal materials. According to the relevant statistics, there are nearly 300 varieties of common species which are more than one million kilograms. Some traditional medicines were very popular at home and abroad and had good sales before the founding of the People's Republic of China, such as Deqing fleece-flower root, Lianzhou *radix polygonati officinalis*, Xijiang *cinnamon*, Heyuan *ficus simplicissima lour*, Xinhui *tangerine peel*, Shipai *agastache rugosus*, Huazhou *tangerine*, Yangchun *fructus amomi*, *Citrus medica*, *Morinda officinalis* How, *Aquilaria sinensis*, *Alpinia offcinarum*, *Camellia petelotii*, Guangxi *clerodendrum cyrtophyllum*, *Pheretima*, *Bungarus parvus*, etc. In addition, Cinnamomum, *Santalum album*, *Ginkgo biloba*, *Eucommia ulmoides*,

Phellodendron amurense, *Magnolia officinalis*, *Evodia rutaecarpa*, *Vitex trifolia*, *Dioscorea polystachya*, *Alisma plantagoaquatica*, Trichosanthes, *Quispualis indica*, *Andrographis paniculata*, *Perilla frutescens*, *Desmodium styracifolium*, *Lythrum salicaria*, *Mentha canadensis*, *Typhonium cuspidatum*, *Zigiber officinale*, *Curcma longa*, *Kaempferia galanga*, *Amomum*, *Polygonatum Sibiricum*, *Sauropus spatulifolius*, *Aloe vera* Var. *chinensis*, *Wolfiporia extensa*, *Ganoderma lucidum* are common family medicinal herbs. Lingnan wild plants mainly include: *Aquilaria sinensis*, *Lophatherum gracile*, *Lonicera confusa*, *Smilax glabra*, *Litsea pungens*, *Lygodium japonicum*, *Brucea javanica*, *Imperate cylindrica* Var. *major*, *Helicteres angustifolia*, *Anisomeles indica*, *Ilex pubescens*, *Rosa laevigata*, *Houttuynia cordata*, *Cyperus rotundus*, *Bidens pilosa*, *Juncus effusus*, *Kadsura longipedunculata*, etc. Wild and bred medicinal animals include: Gecko, Leech, Pherefima, Scorpion, Sands cattle, Oyster, Silkworm. There are also a large number of marine medicinal organisms such as Seahorses, Sea dragons, Sea stars, Algae, Partridges, and Shellfish.

Guangdong is the famous province for traditional Chinese medicine. According to the research, there are more than 4, 500 medicinal resources in Lingnan area, accounting for 36% of the national medicinal resources. Among them, Guangdong province produces 20.7% of the country's medicinal materials, while the total number of medicinal herbs in Guangxi ranks the second, behind Yunnan. According to the data of Chinese traditional Chinese medicine resources survey in the 1980s, there were 2, 614 kinds of traditional Chinese medicine resources, of which 2, 500 species of medicinal plants belong to 225 families, 1, 175 generas; it has 109 species medicinal animals, belonging to 89 families; 25 kinds of medicinal minerals. At that time, there were 1, 170 kinds of medicinal plants, which were compiled into the catalogue of *Guangdong Traditional Chinese Medicine Resources*. There are 109 kinds of medicinal animals, and 24 kinds of medicinal minerals. The total number is 1, 303. There are 548 families of Chinese traditional medicine in Guangxi Zhuang Autonomous Region, 1, 861 genus, 4, 637 species (including subspecies, varieties and variants) , of which algae has 12 families, 12 generas, 15 species; fungi has 28 families, 49 generas, 85 species; lichens has 5 families, 7 generas, 10 species; bryophyte has 12 families, 13 generas, 15 species; furn has 46 families, 89 generas, 225 species; medical plant has 221 families, 1, 343 generas, 3, 728 species; medical animal has 214 families, 348 generas, 509 species; mineral has 50 species. In addition, abundant germplasm resources of traditional Chinese medicine provide a strong material base for the formation and development of local traditional Chinese medicine industry. There are nearly 10 famous enterprises in Guangdong province with a history of more than 100 years, such as Chen Liji, Feng Liaoxing, Zhong Yi Pharmaceutical Corporation, Jing Xiutang Pharmaceutical Corporation, Wang Laoji, Pan Gaoshou Pharmaceutical Corporation, etc.

1. The location of LIngnan traditional Chinese medicine resources

(1) Traditional Chinese medicine plants

The growth and development of plants is closely related to the ecological environment. There is no cold winter in Lingnan area, but the climate, soil and vegetation shows distinct zonal distribution. Therefore, the traditional Chinese medicine also has zonal distribution. The different kinds of traditional Chinese medicine in Lingnan is stated in the following according to their geographical location and natural conditions.

① Tropical region

The southwest of Guangdong Province, the southeast of Guangxi Zhuang Autonomous Region, Hainan Province are the tropical region, where the accumulated temperature is from 8,000 ℃ to 8,500 ℃. There are long summers and mild winters. The yearly rainfall is from 1,500 mm to 2,500 mm, making it a tropical monsoon climate. The soil in the region is typically brick red soil. The vegetation types are tropical rainforest, tropical monsoon forest, mangrove, tropical grasslands, and tropical beach sand plant.

The types of traditional Chinese medicine grow differently in the vertical distribution because of the different landforms and vegetation. The sea level of Leizhou Peninsula, which is a basalt table land, is below 100 m and the land surface fluctuates slightly. There are mainly low terraces, sandy beaches and mangrove belts in this region. The main medicine plants resources include Dracaena draco, *Cinnamomum cassia* Presl, *Barbados aloe, Alpinia oxy phyllamiq., Areca catechu* Linn., Stephania succifera, Japanese stephania root, etc.

② Low terrace

There is a large population here. The original vegetation almost disappeared because of long-term development and utilization. Most of the vegetation now is cultivated plants. The traditional Chinese medicine resources of wild plants here include: *Wisktroemia indica, Clerodendrum cyrtophyllum, Helicteres angustifolia, Asparagus fern, Flemingia prostrata, Millettia speciosa, Croton crassifolius, Microcos paniculata, Rhodomyrtus tomentosa, Ilex rotunda, Tadehagi triquetrum, Desmodium styracifolium, Smilax glabra, Lygodium japonicum, Brucea javanica, Achyranthes aspera, Verbena officinalis, Imperate cylindrica* var. *major, Alpinia officinarum, Canavalia maritima, Maclura tricuspidata Carrière, Citrus medica, Elsholtzia ciliata, Striga asiatica, Evolvulus alsinoides, Adenosma indianum, Cyperus rotundus, Centella asiatica, Herba elephantopi, Bidens pilosa*, etc. The planted traditional Chinese medicines include *Pogostemon cablin, Alpinia officinarum, Kaempferia galanga, Andrographis paniculata* and *Lythrum salicaria*.

③ Sandy beach

This region is located along the coast. On average, the region is 10 m above sea level. It is hot and dry and has intense solar radiation. It is high in soil salinity because of the influence of sea winds and seawater. Vegetation in this area includes *Vitex rotundifolia, Cyperus rotundus, Ipomoea pes-caprae, Canavalia maritima, Alove vera* var. *Chinensis, Catharanthus roseus, Opuntia dillenii*, and *Asparagus cochinchinensis*.

④ Mangrove belt

This is a special ecological type near tropical beaches. The vegetation here is mangrove and semi-mangrove. The plant resources are mainly *Clerodendrum inerme, Bruguiera gymnorhiza, Ceriops tagal, Avicennia marina, Pandanus cerberamanghas,* and *Scaevola sericea*.

⑤ South subtropical region

This region is located east of the most tropical regions of China, south of Huaiji, Qingyuan, Fogang, Longchuan, and Dapu in Guangdong Province. The southeast part of this region is near the sea and the southwest part of this region is next to the tropical vegetation zone in the northern part of the Leizhou Peninsula. The topography of this region is mainly mountains and hills and the other part is delta alluvial plains.

This area belongs to the tropical monsoon type and has observable dry and wet seasons, with the tropic of cancer running through its northern part. The yearly average temperature is from 20 °C to 22 °C. The lowest temperature is 0 °C in cold climate with light frost. The yearly rainfall here is from 1,500 mm to 2,200 mm. The soil is lateritic soil, mountainous red soil and mountainous yellow soil. The type of the vegetation is subtropical evergreen, monsoon forest, secondary subtropical grass slopes and plantation.

The vegetation is tropical and subtropical transitional types and the tropical floristic elements account for the majority of species. There are many varieties of wild and domesticated traditional Chinese medicines. The planted ones include: *Amomum villosum, Fallopia multiflora, Citrus maxima, Ailanthus altissima, wolfiporia extensa, Alisma plantago-aquatica, Pogostemon cablin, Citrus medica, Perilla frutescens, Rhizoma zingiber officinale, Radix curcumate, Kaempferia galanga, Lonicera japonica* var. *chinensis, Belamcanda chinensis, Homalomena occulta, Euryale ferox, Andrographis paniculata*, etc.

The wild ones include: *Momordica cochinchinensis, Vites trifolia, Alpinia katsumadai, Croton tiglium, Abrus precatorius, Pueraria montana, Viola philippica, Aristolochia fangchi, Lycopodium clavatum, Croton crassifolius, Mahonia fortunei, Desmodium styracifolium, Schefflera heptaphylla* and *Prunella vulgaris*, etc.

⑥ Mid-subtropical region

Mountains and hills are the dominant terrain in the city of Qingyuan, Shaoguan, Heyuan, Meizhou, which are in the northern part of Guangdong province. The climate changes a great deal in a calendar year. The average temperature is from 18 °C to 20 °C. The lowest temperature is below 0 °C. The frost season is about one and half months, with freezing and snowy conditions. The yearly rainfall is around 1,500 mm and it is from 1,700 mm to 2,000 mm for some regions. The spring rain falls early and the rainfall season is distributed evenly. The climate here is wet with a short dry season. This region has the most various types of wild traditional Chinese medicine resources.

The vegetation here is mainly subtropical flora and as well as tropical flora and mountain types. There is also temperate flora in the region. The planted traditional

Chinese herbs include: Magnolia officinalis, *Eucommia ulmoides, Panax notoginseng, Polygonatum odoratum, Platycodon grandiflorus, Atractylodes macrocephala, Angelica dahurica*, etc. The wild types include *Aristolochia debilis, Polygala Lenuifolia, Velutinous cinquefoil root, Ligustrum Lucidum, Serratula chinensis, Salvia bowleyana* and *Polygonatum sibiricum*.

There is a big difference in vegetation in vertical distribution. The types of traditional Chinese medicine in mountains include: *Coptis chinensis, Panax notoginseng, Cephalotaxus fortunei, Gentian, Ligusticum sinense, Arisaema heterophyllum, Polygonatum sibiricum, Asarum insigne* and *Uncaria rhynchophylla*. The types of traditional Chinese medicine in open forests and channel edge contain: *Strobilanthes cusia, Lignum millettiae, Cyrtomium fortunei; Polygonum cuspidatum, Rubia cordifolia, Rabdosia serra, Akebia quinata* and *Trichosanthes uniflora*, etc. The types of medicine in the hills consist of *Carculigo orchioides, Litsea cubeba, Lindera aggregata, Wikstroemia indica, Ilex pubescens, Roughhaired holly root, Fructus gardeniae, Lonicera japonica, Rosa laevigata, Kadsura longepedunculata, Kadsura coccinea, Pueraria montana* var. *lobata* and *Smilax glabra*, etc. The most well-known traditional Chinese medicine plants on the grassy slope are *Xanthium strumarium, Helicters angustifolia, Viola philippica, Striga asiatica, Buchnera* and *Eupatorium*, etc. The traditional Chinese medicine plants on the region's plains and fields are *Hibiscus syriacus, Cinnamomum camphora, Morus alba, Melia azedarach, Hibiscus mutabilis, Citrus medica, Chrysanthemum morifolium, Mentha canadensis, Leonurus japonicas, Artemisia argyi, Herba ecliptae, Grangea maderaspatana, Herba plantaginis, Hedyotis diffusa, Houttuynia cordata, Prunella vulgaris, Lobelia chinensis*, etc. The plants in river, lake and everglade are Nelumbo nucifera, Reed rhizome and Acorus calamus.

(2) Traditional Chinese medicine of animals and mineral

① Medicinal animal medicine

Terrestrial animals include: Panthera tigris amoyensis, Cervus nippon, River deer, Monkey, Gecko, Fossilia spiriferis, Tortoise, Trionyx sinensis, Manis pentadactyla, Leopard, Snake, Pberetima, Honeycomb, Bat, Cicada, Myrmeleontid larva, Carpenter bee and mantis, Toad. Marine organisms include: Coral, Abalone, Concha cypraeae alba, Concha arcae, Pearl shell, Limulus polyphemus, Starfish, Sea urchin, Shark, Hippocampus, Sea dragon, Sea sparrow, Sea turtle, Hawksbill turtle and Sea serpent. There are also various types of seaweed and digenia.

② Medicinal mineral

Here we take Guangdong Province as an example, see table 1 the major mineral medicines and their distirbution.

Table 1. Guangdong Province mineral medicines and their distribution

Traditional Chinese Medicine Name	Major Distribution Districts/Counties
Pumex	Zhanjiang, Zhuhai, Dianbai, Taishan, Yangjiang, Huilai, Chaoyang, Huidong, Haifeng, Lufeng, etc.
Ophicalcitum	Heping
Ochre	Shunde
Stone of Sulfur	Yingde
Gypsum	Xingning
Gypsum Rubrum	Qingyuan, Qujiang, Shixing, Heping, Lianjiang, Gaozhou, Xinyi, Jiaoling, Meixian, etc.
Limonitum	Qingyuan
Talc	Liannan, Gaozhou, Xinyi, Yangchuan, Lianjiang, etc.
Magnetite	Dapu, Yangchun, Heping, Xinfeng, Fogang, etc.
Pyrite	Lechang, Yangchun, Kaiping, Enping, Xingning, etc.
Stalactite	Meixian, Jiaoling, Yangchun, Huaiji, Fengkai, Shaoguan, Huidong, Heping, Haifeng, Qingyuan, etc.
Purple White Quartz	Qingyuan, Heyuan, etc.

Part of mineral medicine: Stalactitum, Reddle, Steatite, Amethgst, Halloy situm rubrum

2. Traditional Chinese medicine resource distribution areas of Guangdong Province

Guangdong is divided into six resource distribution areas according to natural geography characteristics, resource construction, distribution, and traditional Chinese medicine production history. There are other professional elements such as agriculture, forestry, soil, climate and vegetation.

(1) Northern Guangdong contains medicinal areas within the mountains and hills in northeast region

This area accounts for one third of the total land mass of Guangdong. It has abundant traditional Chinese medicine resources and has medicine both from northern and southern areas. It has various kinds of medicine, among which animal resources ranks first. Herbal medicines include: *Long-noded pit viper, Scorpio, Centipede, Leech, Pherefima* and *Ground beetle*. The major planted medicines include: *Eucommia ulmoides, Magnolia officinalis, Gingko biloba, Polygonatum odoratum, Polygonatum sibiricum,* and *Wolfiporia extensa*, etc. Wild resources include: *Selaginella tamariscina, Woodwardia japonica, Cyrtomium fortunei, Lindera aggregata, Pueraria montana, Rosa laevigata, Holly root, Crataegus Pinnatifida, Lonicera japonica* var. *chinensis, Lsodon serra,* and *Lophatherum gracile*.

(2) Medicinal areas of hills and tablelands in southeast Guangdong

This area has various landscapes and tablelands. The northern-most terrain has a higher elevation. The area faces the sea with hills for its background. Plains and shallow tidal flats account for a certain amount of area which are suitable for all kinds of herbal medicines growth. The planted herbal medicines include: *Andrographis paniculata, Desmodium styracifolium, Lsodon serra, Aquilaria sinensis, Kaempferia galange,* and *Alpinia officinarum*. This area has a history of introducing herbs from other provinces, and the major varieties include: *Codonopsis pilosula, Ligusticum wallichii,* and *Rehmannia glufinosa*.

(3) Medicinal materials of Pearl River Delta

There has been a very rapid economic development of this area within the last ten years, but the planted herbs have relatively decreased. The traditional herbs of this area include: *Anisochilus carnosus, Juncus effusus Perilla frutescens, Jasminum grandiflorum,* and *Lycianthes biflora* (lourei) *bitter*. Wild resources include: *Rhododendron emesiae, Cibotium barometz, Sargentodoxa cuneata, Davallia mariesii, Imperata cylindrica, Microcos paniculata, Smilax glabra,* and *Helicteres angustifolia*, etc.

(4) Medicinal areas of hills and mountains in western Guangdong

This area has many Guangdong specialty medicinal materials and is the major production area of medicinal materials within the province. The major varieties include: *Citrus reticulata, Amomum villosum, Citrus maxima, citrus medica, Morinda officinalis Cinnamomum cassia, Pogostemon cablin, Fallopia multiflora, Lonicera japonica* var. *chinensis, sauropus spatulifolius, Euryale ferox, curcuma aromatica, Curcuma longa, Illicium verum, Aquilaria sinensis, satalum album, Alpinia oxyphylla, Dioscorea polystachya, Wolfiporia extensa* and *Pueraria montana* var. *thomsonii*. It has abundant wild resources.

(5) Tropical medicinal areas of Leizhou peninsula

This area is the tropical peninsula of the southwest China mainland. It has abundant heat resources and suitable for developing lots of tropical medicines. It exerts some influences due to the drought and lack of water. The major planted herbs include: *Andrographis paniculata, Desmodium styracifolium* and *Lythrum salicaria*. Zhanjiang south medicine experiment area of the north town in Suixi ridge has successfully introduced many imported medicines since the 1960s. *Santalum album* is the major herb, and it also has other herbs like *Acacia catechu, Hydnocarpus anthelminticus, strychnos nux-vomica, Styrax benzoin*, and *Terminalia chebula*.

(6) Sea marine medicinal areas in the south China Sea

The coastline of the province is 3,368.1 km, ranking first in China. It has abundant marine herbal medicines. In Haifeng, it has collection of 108 pieces of marine herbal medicine specimens. The major varieties include: Sea horse, Sea otter, Hawksbill, Oysters, Seaweed, Lamsbarua haoibsca aresh, Ulva, and Gracilaria. The mari-culture varieties include: Sea horse, Pearl and Abalone.

3. Lingnan major specialty herbal medicines and their production status

(1) *Pogostemon cablin* (Blanco) Benth.

It is Lamiaceae *Pogostemon cablin* (Blanco) Benth. with grass in medicine. It is pungent in flavor and warm in property. It effects spleen, stomach and lung properties. It has the function of aroma-therapy, stimulate appetite, prevent nausea, and relieve summer-heat. It is used to cure the retention of damp-turbid substances, abdominal distension, vomiting, summer heat, fatigue, chest congestion, coldness, dampness and heat, abdominal pain, diarrhea, nasosinusitis, and headaches. Patchouli is the raw material of many Chinese patent medicines. Patchouli oil is the major material of the pharmaceutical industry, light industry (spices, essence, and perfume), also it is an important export product.

Patchouli was originally produced in India and Malaysia. Guangdong, Guangxi, Yunnan, Hainan province planted it in the early Song Dynasty. In the past, Guangzhou and Zhaoqing were the two major production cities with fine quality. Now cities like Zhanjiang, Yangjiang, Suixi, Wuchuan and Yangchun of Maoming produce the most. Cities and counties like Dianbai, Huazhou, Xuwen, Lianjiang, Sihui, Gaoyao also plant it.

(2) *Citrus medica* L.var. *sarcodactylis* Swingle

It is the fruit of rutaceae *Citrus medica* L.var. *sarcodactylis* Swingle. It has a pungent, bitter, acidic flavor, and warming properties. It belongs to the liver and stomach meridian. It has the function of soothing the Qi of liver and the collateral of dredging, stomach relief, and prevention pain. It is used to cure stagnation of Qi in the liver and stomach, fullness in chest, ribcage, and stomach, poor appetite, vomiting and dispelling the effects of alcohol.

Citri medica is mainly located in the cities and counties like Gaoyao, Sihui, Deqing and Yunfu. Other places like Heyuan in eastern Guangdong, Meizhou, Chaozhou, Lianjiang in western Guangdong also plant it in recent years.

(3) *Santalum album* Linn.

It is Santalaceae *Santalum ablum* Linn., which is used for heart treatments. It has the function of promoting circulation of Qi and helping keep warm, stimulating appetite, and pain relief. It is used to cure stagnation of Qi which is caused by the pathogenic cold, chest pain, abdominal pain, stomachache, appetite loss, coronary heart disease, and angina. The wood of sandalwood is aromatic and solid, which is suitable for making sandalwood carvings. Sandalwood oil has a precious fragrance which attracts a high price.

Sandalwood is mainly produced in India, Indonesia and Malaysia. Guangdong has started to plant this herb since 1962. With thirty years of experimentation and promotion, Chinese scientists have concluded an applicable plant technology regarding this plant. Now cities and counties like Yangxi, Deqing, Lianjiang, Huazhou, Suixi, Dianba, Guangzhou and the surrounding areas have planted this herb.

(4) *Morinda officinalis* How

Its formal name is *Rubiaceae morinda officinalis*. It has a pungent sweet flavor, and has warming properties. It effects the liver and kidney. It has the function of promoting kidney health, strengthening bones and musculature, and alleviating bloating. It is also used to cure erectile dysfunction, lower abdomen pain, irregular menstruation, deficiency cold of the uterus, wind-cold-dampness arthralgia, aching limbs and knees.

Morinda officinalis is mainly produced in Guangdong, Guangxi, and Vietnam. Major producing cities and counties are Gaoyao, Deqing, Yunan, Huaiji, Wuhua, Heyuan; Gaoyao in Guangdong Province and Deqing produces the most.

(5) *South Cinnamomun cassia* Presl

It is a modification of the Lauraceae *Cinnamomum cassia* Presl. At first, it was called Vietnam thanh hoa cinnamon, and was named South Cinnamon temporarily. The bark, twig (cassia twig), fruit are used as medicine. Bark: Strengthening the Yang Qi while warminging the spleen and stomach, alleviating cold symptoms, and promoting blood calculation. Twig: Presenting and relieving the skin, warming and circulating the venation. Fruit: Warming the stomach.

South cinnamon is mainly produced in cities and counties like Xinyi, Gaoyao, Luoding, Deqing, Yunan, Yunfu. The plant area of Guangdong province contained over 140 hectares until the end of 1986. The plant area of Xinyi is the largest one. Gaoyao, Luoding, Deqing and areas along the Xijiang river mainly produce Xijiang cinnamon.

(6) *Amomurn villosum* Lour.

Zingiberaceae *Amomum villosum* lour. It is the main ingredient of the traditional Chinese medicine Amomum villosum. It is pungent in flavor and warm in property. It belongs to spleen, stomach and kidney meridian. It has the function of resolving dampness, stimulating appetite, warming spleen, preventing diarrhea, preventing vomit, calculating Qi, and preventing miscarriage. It is used to cure retention of damp and resistance, gastric fullness, loss of appetite, deficient cold in spleen and stomach, pregnancy instability, and threaten abortion.

Fructus amomi is our specialty traditional Chinese medicine. Yangchun is the famous production place, and other cities like Yunnan, Hainan, Xinyi, Gaozhou, Guangning, Fengkai are also production places.

(7) *Alpinia officinarum* Hance

It is the dry rhizome of zingiberaceae plants *Alpinia officinarum* Hance. It is pungent in flavor and hot in property. It belongs to spleen and stomach meridian. It has the function of warming the stomach and dispelling cold as well as pain relief and promoting digestion. It is used to cure epigastria distension, cold symptoms, stomachache, vomiting, and belching.

The root of Galangal is mainly produced in Xuwen, Leizhou, Suixi and Lianjiang. Countries like Longtang, Fucheng, Qujie, Qianshan, Jinhe of Suixi have large planted areas with good qualities.

(8) *Kaempferia galanga* Linn.

Zingiberaceae *Kaempferia galanga* Linn.. It is pungent in flavor and warming in property. It belongs to stomach meridian. It has the function of keeping warm, promoting digestion, and pain relief. It is used to cure chest or diaphragm distension, epigastria distension, cold pain, dyspepsia, traumatic injury, and toothaches. Besides, Kaempferia galangal is also used as a seasoning material for several subsidiary foods.

Kaempferia galangal has a wide range of plant areas in Guangdong province. It is planted in cities like Meizhou, Heyuan, Huizhou, Chaozhou, Zhanjiang, Maoming, Yangjiang and Zhaoqing.

(9) *Citrus maxima* 'Tomentosa'

Rutaceae Citrus granolis (L.), belongs to spleen meridian. It has the function of dispelling cold, eliminating dampness, promoting circulation of Qi, stopping coughing, and removing phlegm. It is used to cure coughing, itchy throats, excessive phlegm, dyspepsia, vomiting, and chest pains. It has over 20 pharmaceutical factories which use exocarpium as raw material to produce Chinese patented medicines. The exocarpiums demand exceeds supply and some factories use shaddock peel to replace which affects the quality of certain medicines.

Exocarpium is mainly produced in Huazhou and it has hundreds of years of plant history. Other places like Lianjiang, Suixi and Maoming also produce it.

However, Huazhou is known to produce the best quality.

(10) *Citrus reticulata* Blanco

It is Rutaceae *Citrus chachiensis* Hoet., *C. tankan* Hayata., with mature fruit. It is pungent in flavor and warm in property. It belongs to lung and spleen meridian. It has the function of prompting circulation of Qi and spleen, eliminating dampness and phlegm. It is used to cure chest and diaphragm fullness and distention, dyspepsia, vomit, cough and excessive phlegm.

Pericarpium citri reticulatae is one of the "Top Ten Guangdong Medicines" and its major producing areas include Xinhui, Sihui, Chaozhou, Boluo and Puning. In Guangdong, Sihui's percarpium citri reticulatae is of good quality and a major export product.

(11) Hippocampus

It is the syngnathidae animal Hippocampus kelleggi Jordan et Sugder, H. histrix kaup, H.kuda bleeket, H. trimaculatus leach or without viscera Japanese H. japonicus kaup. It is sweet in flavor and it has warming properties. It belongs to liver and kidney meridian. It has the function of promoting circulation of Qi and modifying Yang, and improving blood circulation. It is used to cure erectile dysfunction, enuresis, deficient dyspnea, dystocia, furuncle, and poisonous swelling.

Sea horse is produced in the coastal bay of tropical and temperate zones, as well as the low quiet tidal regions, with clear water, rich algae breeding, and zooplankton. Cities and countries like Lufeng, Leizhou, Dianbai, Wuchuan have artificial breeding histories.

(12) Pheretima

It is megascolecidae animal with dry *Pheretima aspergillum* (Perrier), which called "pheretima". It is salty in flavor and cold in property. It belongs to liver, spleen, and bladder meridian. It has the function of alleviating the heat and relieving convulsions, smoothing collaterals, relieving asthma, and diuresis. It is used to cure high fever, dizziness, scarring, arthralgia spasm pain, extremities numbness, hemiplegia, lung pressure, asthmatic cough, prostate problems, edema, and hypertension.

Pheretima is a wild variety and mainly produced in Foshan, Jiangmen, Huizhou, Heyuan, Meizhou, and Maoming. Heshan and Nanhai have a long history of pheretima exportation.

第二章
岭南特产中草药资源

岭南特产药材资源十分丰富，也是岭南中草药的重点品种。据统计，广西壮族自治区特有药用植物有112种，如长茎金耳环、广西大青、细柄买麻藤、广西斑鸠菊、金花茶等；广东省特有药用植物近60种，其中包括著名的十大广药：砂仁、广巴戟、石牌藿香、化州橘红、肇庆佛手、高良姜、新会成皮、沉香、广地龙、金钱白花蛇；海南省区原产药材有10余种，如槟榔、益智仁、巴戟天、砂仁、沉香、银花、山栀子、藿香、降香、红壳松等。其中，砂仁、广巴戟、槟榔、益智仁是我国传统的四大南药。

Chapter 2 Lingnan specialty Chinese herbal medicine resources

Lingnan has abundant specialty medicine resources and it is also the key species of the Lingnan Chinese herbal medicine. According to statistics, there are 112 species of medicinal plants in Guangxi Zhuang Autonomous Region, such as the asarum longerhizomatosum, Guangxi clerodendron cyrtophyllum turcz, gnetum gracilipes, vernonia chingiana, golden camellia, etc. There are nearly 60 special medicine plants in Guangdong Province, which include 10 famous Guangdong medicines: Yangchun *Amomum villosum Lour.*, Guangdong *Radix Morindae officinalis*, Shipai *agastache rugosus*, *Citrus maxima*, Zhaoqing *fructus citrus sarcodactylis*, *Alpinia officinarum*, Xinhui *pericarpium citri reticulatae*, *Aquilaria sinensis*, *pheretima*, coin-like white-banded snake bungarus parvus. There are more than 10 medicinal herbs in Hainan province, such as *Areca catechu*, Alpinia oxyphycca, Morinda officinalis, Amomum villosum, Aquilaria sinensis, Lonicera japanica, Gardenia jasminoides, Agastache rugosus, Dalbergia odorifera, Cephalotaxus sinensis, etc. Among them, the Yangchun *Amomum villosum*, Guangdong *Morinda officinalis*, *Areca catechu* and *Alpinia oxyphylla* are the four traditional southern China medicines.

1. 佛 手

Citrus medica L. var. *sarcodactylis* Swingle

科属：芸香科，柑橘属。

药名：佛手。

别名：广佛手、佛手果、五指柑、十指柑。

植物特征：灌木状，具有短而硬的刺。单叶互生，叶子边缘有波状钝锯齿。花单生、簇生或为总状花序，萼片为5裂片；花瓣5片，子房在花柱脱落后即行分裂。成熟果顶端裂瓣如手指状肉条，果皮甚厚。

资源状况：广佛手主要种植于高要、四会、德庆、云浮等地，近年来粤东的河源市、梅州市、潮州市以及粤西的廉江市也有种植，由于市场供需的关系，价格和种植面积波动较大。

药用部位：果实。

功能主治：理气止痛，消食化痰。可治胸腹胀满、食欲不振、胃痛等。

主要成分：果实含柠檬油素、地奥明、橙皮柑等。

Genus: Rutaceae, Citrus.

Scientific name: *Citrus medlica* L. var. *sarcodactylis* Swingle.

English Name: Fructus citri sarcodactylis.

Features: Small trees or shrubs, with short and hard spiny branches. The flowers are solitary, fascicled, or raceme with 5-lobed sepals and 5 petals. The ovary is divided after the style falls off. The yellow mature fruit splits like a fist or opens as a finger with coarse and thick skin. The florescene is from April to May. The fruit period is from October to December.

Resources Situation: *Citrus meclica* L.var. *sarcodactylis* Swingle is mostly planted in Gaoyao, Sihui, Deqing and Yunfu. It has been planted in Heyuan, Meizhou, Chaozhou and Lianjiang. Due to the market supply and demand, the price and the planting areas have changed frequently.

Medicinal Part: The fruit is considered the most medicinal part of the plant.

Functions: Strengthening spleens and stomachs, circulating tendons, detoxicating lungs, dissolving phlegm, and relieving pain.

Main Ingredients: Citropten, diosmin, hesperiden.

2. 广藿香

***Pogostemon cablin* (Blanco) Benth.**

科属：唇形科，刺蕊草属。

药名：广藿香。

别名：刺香、刺蕊草、排香草、大叶薄荷。

植物特征：多年生草本或半灌木，叶子两面均被茸毛。叶对生，揉搓之有清淡的特异香气；轮伞花序密集，花冠淡红紫色，花冠筒伸出萼外，冠檐二唇形。广藿香在广东极少开花。花期4月。

资源状况：广藿香原产于印度和马来西亚，我国岭南（今广东省）在宋代或更早已有引种，广西壮族自治区、云南省、海南省也有栽培。过去广州市、肇庆市是主要产地，而且质量最好，现在主产地是湛江市、阳江市、茂名市，以遂溪县、吴川市、阳春市最多。广州的石牌藿香由于城市发展，目前只有少量作为试验保护，无法供应市场。

药用部位：全草。

功能主治：芳香化浊，开胃止呕，发表解暑。可治湿浊中阻、脘痞呕吐、暑湿倦怠、胸闷不舒等。属化湿药。

主要成分：含挥发油，主要成分为广藿香醇、广藿香酮等。

Genus: Labiatae, Pogostemon.

Scientific Name: *Pogostemon cablin* (Blanco) Benth..

English Name: Herba pogostemonis.

Features: Perennial herb or subshrub with hair on both sides of the leaves. The leaves alternate and are malaxate with light-favor. It has a dense verticillate light, red-purple crown. Corolla tube extends to the leaf and the crown edge is bilabiate. Patchouli is rarely bloom in Guangdong. The florescence is in April.

Resources Situation: *Pogostemon cablin* (Blanco) Benth is mostly produced in India and Malaysia. It was planted in Lingnan (Guangdong Province) in Song dynasty and earlier dynasty, and Guangxi Zhuang Autonomous Region, Yunnan Province and Hainan Province also planted it before. In the past, Guangzhou and Zhaoqing were the two major producing cities with high quality. Now the major production cities are Zhanjiang, Yangjiang, Maoming, Suixi, Wuchuan and Yangchun. Except for a small number of it for experiment protection, most local brand patchouli cannot supply the market due to the city's development.

Medicinal Part: The whole herb.

Functions: Eliminating turbid pathogen with aromatics, stimulating appetite, preventing vomit, relieving heat rash. It is also known for curing retention of damp-turbid substance, tiredness, and chest distress.

Main Ingredients: It contains volatile oil but the main ingredients are patchouli alcohol and patchoulenone.

3. 春砂仁

Amomum villosum Lour.

科属：姜科，豆蔻属。

药名：春砂仁。

别名：阳春砂、小豆蔻。

植物特征：多年生草本，根状茎圆柱形，匍匐地面，节上有筒状、紫褐色膜质鳞片。叶鞘开放，抱茎，叶舌膜质，根状茎抽出的球状、松散穗状花序，每穗有花7~13朵，萼白色，顶端3浅裂，花冠白色，圆匙形，果实初为绿色，后渐变为红色至紫红色，球形或卵圆形，外面有柔刺。花期3~6月，果期7~9月。

资源状况：春砂仁是我国特产中药材，历年来广东省阳春市出产最为地道，云南省、海南省也有种植，广东省除阳春外还有信宜、高州、广宁、封开等均有产。

药用部位：干燥成熟果实。

功能主治：行气化湿，温脾止泻，理气安胎。用于治疗脘腹胀痛、呕吐泻泄、胎动不安等。

主要成分：主含挥发油，油中主要成分为龙脑，其次为樟脑、乙酸龙脑酯等。

Genus: Zingiberaceae, Amomum.

Scientific Name: *Amomum villosum* Lour..

English Name: Fructus amomi.

Features: Perennial herb with cylindric rhizomes. It features prostrate with tubular on the the base and purple-brown membranous scales on the knob. The stem has open blade tips. It also has ligule membranous and the rhizome drawn balls and loose spikes with 7-13 flowers per spike. It appears with a white color and at the top of the spire has three super-sulcuses with white corollas which are shaped like round spoons. The fruit is green, spherical or oval, and then turns red or purplish-red with soft spines outside. The florescence is from March to Jane, and the fruit period is from July to September.

Resources Situation: *Amomum villosum* Lour is a Chinese specialty medicine. Yangchun of Guangdong Province produces the most famous plants. Provinces like Yunnan, Hainan, cities like Xinyi, Gaozhou, Guangning, Fengkai in Guangdong also produce it.

Medicinal Part: The dried mature fruit is considered the most medicinal part of the plant.

Functions: Promoting circulation of Qi, resolving excessive fluids, warming spleen and stopping diarrha, preventing miscarriage, relieving abdominal distention, vomiting, and nausea.

Main Ingredients: Volatile oil, main ingredients are bronyl acetate, camphor, borneol saponin.

4. 巴戟天

***Morinda officinalis* How**

科属：茜草科，巴戟天属。

药名：巴戟天。

别名：鸡肠风、兔子肠、巴吉天。

植物特征：藤状灌木。根肉质肥厚，圆柱形。叶对生，全缘，小枝幼时中脉上被短粗毛，后变粗糙，叶柄有褐色粗毛；托叶鞘状。花序头状，生于小枝顶端；花萼倒圆锥状，花冠肉质白色。浆果近似球状。花期4~7月，果期6~11月。

资源状况：巴戟天主产于广东省、广西壮族自治区、福建省等地，越南也有产，广东省主要产于高要、德庆、郁南、怀集、五华、河源等市县，以高要、德庆产量最大。属国家三级濒危保护植物。

药用部位：干燥根。

功能主治：补肾阳，壮筋骨，祛风湿。可治阳痿、少腹冷痛、小便不禁、子宫虚冷等。属补虚药下属分类的补阳药。

Genus: Rubiaceae, Morinda.

Scientific Name: *Morinda officinalis* How.

English Name: Radix morindae officinalis.

Features: It is a twining shrub with round and hypertrop hy fleshy roots. The leaves are opposite with entire margins. The midrib of young branches has a short shag. The petiole has brown shag and the stipule is scabbard-like. Inflorescence is a head shape and in the top branch. The calyx is obconical and the corolla flesh is white. The Baccate is like sphere. The florescence is from April to July. The fruit period is from June to November.

Resources Situation: *Morinda officinalis* How is mainly produced in Guangdong, Guangxi Zhuang Autonomous Region, Fujian province, and Vietnam. In Guangdong, it is mainly produced in Gaoyao, Deqing, Yunan, Huaiji, Wuhua and Heyuan, and it produces the most in Gaoyao, Deqing. It belongs to the third-class national endangered protection plant.

Medicinal Part: The dry root is considered the most medicinal part of the plant.

Functions: Recuperating the kidney functions, strengthening bone and tendons, and alleviating rheumatism. It also cures erectile dysfunction, abdomen pains, incontinence of urine, and regulates uterus temperature. It belongs to Yang-tonifying medicine.

5. 沉 香

***Aquilaria sinensis* (Lour.) Spreng.**

科属：瑞香科，沉香属。

药名：沉香。

别名：牙香树、女儿香。

植物特征：常绿乔木。树皮有坚韧的纤维。单叶互生，叶片两面光滑无毛。伞形花序生于枝顶或叶腋，花被黄绿色，钟形；花被5裂。蒴果木质，倒卵形，被黄褐色短柔毛，成熟时2瓣裂。种子1~2粒，黑褐色，卵形，基部有尾状附属物。花期4月，果期6~7月。

资源状况：主产于广东省沿海的汕尾、湛江、茂名及珠江三角洲地区，除少数野生资源外均为人工栽培。

药用部位：含树脂的木材。

功能主治：行气止痛，降逆调中，温肾纳气。可治胸腹胀闷作痛、胃寒呕吐、呃逆等。

主要成分：主含挥发油，油中主要成分为沉香螺醇、白木香酸、白木香醛等。

Genus: Thymelaeaceae, Aquilaria.

Scientific Name: *Aquilaria sinensis* (Lour.) Spreng..

English Name: Chinese eaglewood.

Features: It is an Aiphyllium. The bark has tough fibers. The leaves alternate and the dual blades are smooth and glabrous. Inflorescence is a shed shape and is in the top branch or leaf axil. The perianth is a greenish-yellow with campaniform shape and has 5 lobes. It has capsule-type wood and is obovate. It has yellow brown pubescence and it is 2-lobed when it is mature. It has 1-2 seeds, black or brown in ovoid shape, and it has caudate appendage at its base. The florescence is in April and the fruit period is from June to July.

Resources Situation: *Aquilaria sinensis* (Lour.) Spreng. is mainly produced in Shanwei, Zhanjiang, Maoming and some pearl river delta areas of Guangdong Province. Except for a small number of wild resources, most aquilaria sinensis are artificial cultivation.

Medicinal Part: Wood with resin is considered the most medicinal part of the plant.

Functions: Promoting circulation of Qi reliving pain, keeping kidney warm, relieving asthma, alleviating stuffy thoracic abdominal distension and pain, stomach cold and vomit, and hiccups.

Main Ingredients: Volatile oil, main ingredients are agarospirol, agarospiric acid, baimuxinal.

6. 高良姜

Alpinia officinarum Hance

科属：姜科，山姜属。

药名：高良姜。

别名：良姜、小良姜、海良姜。

植物特征：多年生草本，根状茎圆柱形。叶2列，叶片线状，顶端尾尖，叶舌膜质，披针形，不2裂。总状花序顶生，花萼管顶端3齿裂；花冠管较萼管稍短，蒴果球形，熟时橘红色。花期4~10月，果期9~11月。

资源状况：高良姜主产于徐闻、雷州、遂溪、廉江，以徐闻县的龙塘、附城、曲介、前山、锦和等乡镇种植面积较大，质量较好。

药用部位：干燥根茎。

功能主治：温中散寒止痛。用于寒凝气滞所致的脘腹冷痛、呕吐泄泻、噎膈反胃。

主要成分：根茎含挥发油，另含黄酮类化合物，主要为高良姜素、山柰酚、槲皮素等。

Genus: Zingiberaceae, Alpinia.

Scientific Name: *Alpinia officinarum* Hance.

English Name: Rhizoma alpiniae officinarum.

Features: Perennial herbs with cylindrical rhizomes. It has 2 rows of linear leaves, the top of which is pointed. The ligule has soft membranes, the leaves are lanceolate, and it has terminal racemes. The tip of the calyx splits into 3 parts. The corolla tube is shorter than the calyx tube. It has round capsules and the ripe ones are orange. The florescene is from April to October. The fruit period is from September to November.

Resource Situation: *Alpinia officinarum* Hance is mainly produced in Xuwen, Leizhou, Suixi and Lianjiang; mostly in Longtang, Fucheng, Qujie, Qianshan and Jinhe of Xuwen County. The quality of galangal in this region is quite exceptional.

Medicinal Part: Dried rhizome.

Functions: Relieving pain, warming the spleen, and relieving cold-like symptoms in the stomach. It also alleviates cold-like symptoms in the stomach and abdomen such as: vomiting, diarrhea, hiccups and nausea.

Main Ingredients: The rhizome contains volatile oils and flavonoids such as galangin, campherol, and meletin, etc.

7. 化州橘红

***Citrus maxima*' Tomentosa'**

科属：芸香科，柑橘属。

药名：化州橘红。

别名：化州柚、橘红皮、柚皮橘红。

植物特征：常绿乔木。幼枝绿色密被细茸毛。叶互生，宽卵形或卵状椭圆形，有柔毛及透明腺点，单生或为总状花序生于叶腋，花梗密被白柔毛；花萼杯状、4浅裂；花瓣4片、长圆形、白色。柑果近似球形，成熟果呈柠檬黄色，幼果绿色，密被厚绒毛等。果皮与果肉不易剥离，白皮层极厚。瓤囊16瓣，果肉味酸苦。

资源状况：化州橘红主产于广东省化州市，在本地有几百年的种植历史，此外廉江、遂溪、茂名等地亦有产，但以化州产的正毛橘红质量最佳。花期4~5月，果期10~11月。

药用部位：未成熟或近成熟外层果皮。

功能主治：化痰止咳，行气宽胸，消食健胃。治痰多咳嗽、气喘、呕吐呃逆。属理气药。

主要成分：果皮主含挥发油。

Genus: Rutaceae, Citrus.

Scientific Name: *Citrus maxima* 'Tomentosa'.

English Name: Pummelo peel.

Features: Small evergreen tree. Its spout has green dense tomentose. Leaves alternate, which is broadly ovate or elliptic ovate. It has villous and transparent glandular dots. It has solitary leaves or with racemes in axils. Densely white pubescence is on stems. Cup-shapes calyx with 4 lobes. It has 4 white and oblong petals. The fruit is round and lemon yellow for ripe ones and green for young fruit with dense pubescence. The fruit has thick white layer and it is not easy to peel. The fruit has 16 pieces with sour and bitter flavor. The florescence is from April to May. The fruit period is from October to November.

Resources Situation: *Citrus maxima* 'Tomentosa' is mainly produced in Huazhou of Guangdong Province with hundreds of years of domestication history. It is also commonly planted in Lianjiang, Suixi and Maoming. The quality of the ones in Huazhou is considered the best.

Medicinal Part: The outer peel of fruit before full maturity is considered the medicinal part of the plant.

Functions: It resolves phlegm and relieves coughing as well as promotes blood circulation and digestion. It is also used as treatment for asthma and vomiting. It is used to regulate Qi.

Main Ingredients: The peel contains medicinal oils.

8. 何首乌

Fallopia multiflora **(Thunb.) Haraldson**

科属：蓼科，何首乌属。

药名：何首乌。

别名：首乌、紫乌藤、九真藤。

植物特征：多年生草质藤本。叶互生，心形，先端渐尖，基部心形或耳状箭形，托叶膜质，鞘状，抱茎，褐色。圆锥花序，花被绿白色、花瓣状，5深裂，外面3片有翅。瘦果椭圆形，具三棱角，色黑而有光泽。花期6~9月，果期10~11月。

资源状况：我国何首乌主要分布在陕西南部、甘肃南部。华东、华中、华南、四川、云南及贵州均有分布，部分省区有栽培。广东德庆县主产的何首乌药材地道、质量好。

药用部位：干燥块根。

功能主治：生首乌可润肠通便、解疮毒。可用于治疗肠燥便秘、痈疽。

主要成分：主含羟基蒽醌类衍生物、卵磷脂及锌、锰、铜、锶、镍等。

Genus: Polygonaceae, Fallopia.

Scientific Name: *Fallopia multiflora* (Thunb.) Haraldson.

English Name: Radix polygoni multiflori.

Features: Perennial herbaceous vine with brown leaves in a clasping heart-shaped design. The leaf base is heart-shaped and the leaf apex becomes sharper. The stipule is membranous and has a sheath-like structure. It has paniculate flowers with white and green perianths, petal-like, parted into five distinct directions. The outside 3 petals have wings. The fruits are oval with prisms with black and shiny traits. The florescence is from June to September. The fruit period is from October to November.

Resources Situation: *Fallopia multiflora* (Thunb.) Haraldson mainly grows in southern Shanxi and southern Gansu. It is also distributed in East China, Central China, South China, Sichuan Province, Yunnan Province and Guizhou Province. The *Fallopia multiflora* (Thunb.) Haraldson is of high quality which is produced in Deqing, Guangdong.

Medicinal Part: The dried root has the part considered for most medicinal applications.

Functions: The immature radix polygoni multiflori is good for digestion and used to cure sores and carbuncles.

Main Ingredients: Hydroxy anthraquinone derivatives, lecithin, zinc, manganese, bronze, strontium, and nickel, etc.

9. 肉 桂

Cinnamomum cassia Presl

科属：樟科，樟属。

药名：肉桂。

别名：玉桂、牡桂、桂皮。

植物特征：常绿乔木。叶互生或近对生，革质，长椭圆形或椭圆状披针形，前端稍急尖，基部钝圆，边缘稍内卷，全缘，具离基3出脉。腋生或近顶生的圆锥花序，花被6片。果实椭圆形，熟时呈黑紫色。花期6~7月，果期10~12月。

资源状况：南肉桂主要分布于信宜、高要、罗定、德庆、郁南、云浮等市、县，以信宜的种植面积最大。高要、罗定、德庆等西江流域以种植西江肉桂为主。

药用部位：树皮、嫩枝。

功能主治：补元阳，暖脾胃，除积冷，通血脉。可用于治疗胃腹冷痛、虚寒泄泻、肾阳不足等。

主要成分：含挥发油（桂皮油）1%~2%、鞣质等。

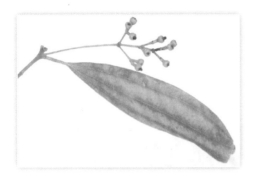

Genus: Lauraceae, Cinnamomum.

Scientific Name: *Cinnamomum cassia* Presl.

English Name: Cortex cinnamomi.

Features: Evergreen trees. The leaves maintain a leathery texture, with alternant or opposite tops, as well as long elliptic or elliptic lanceolate. The leaves have sharp tips and a blunt base with the edge curled inward, containing triple nerved ending. The flowers are panicle, axillary, or terminal. It has a 6-parted flower envelop. The fruits are oval and the mature ones are dark-purple. The florescence is from June to July. The fruit period is from October to December.

Resources Situation: The southern cinnamon mainly grows in Xinyi, Gaoyao, Luoding, Deqing, Yunan and Yunfu countries. There is the maximum production yield in Xinyi. Xijiang areas such as Gaoyao, Luoding and Deqing mainly plant Xijiang Cinnamon.

Medicinal Part: Tree bark and young shoots are considered the most medicinal part of the plant.

Functions: Regulating and nourishing various organs such as the spleen and stomach. It also removes internal cold-like symptoms, and promotes blood circulation. It is often used to cure stomachache, asthenia, diarrhea, and kidney problems.

Main Ingredients: 1%-2% volatile oil (cassia oil), tannin, etc.

10. 广陈皮

Citrus reticulata Blanco

科属：芸香科，柑橘属。
药名：陈皮。
别名：陈皮、橘子皮。
植物特征：小乔木或灌木，少刺。叶片披针形或椭圆形，先端渐尖、微凹。花单朵或2~3朵簇生于叶腋，果扁圆形，果顶略凹，柱痕明显，有时有小脐，蒂部四周有时有放射沟，果皮较薄，油胞明显。花期3~4月，果期11~12月。

资源状况：我国陈皮产地较多，广东主产，其中江门市新会区产的陈皮药材质量最好，传统道地中药称为"广陈皮"，供出口。

药用部位：成熟果皮。

功能主治：理气健脾，燥湿化痰。可治疗脘腹胀痛、食少吐泻、咳嗽痰多等症。

主要成分：果皮含挥发油，挥发油中主要成分为右旋柠檬烯、枸橼醛、α-蒎烯等。

Genus: Rutaceae, Citrus.

Scientific Name: *Citrus reticulata* Blanco.

English Name: Orange peel.

Features: Small trees or shrubs with thorns. The leaves are lanceolate or oval. The top of the leaves often becomes sharper and sharper and have an obvious concave design. Single flower or 2-3 flowers grow in leaf axils. The fruits are oblate. The peel is thin with distinct cells. The florescence is from March to April, the fruit period is from November to December.

Resources Situation: *Citrus reticulata* Blanco grows in different areas in China and Guangdong is the main producing area. The quality of *Citrus reticulata* Blanco in Jiangmen is the best, which is called pericarpium citri reticulatae-chachiensisand and is commonly exported.

Medicinal Part: Peel of mature fruits.

Functions: Regulating blood circulation and strengthening spleen functions, eliminating excess water and phlegm, curing abdominal fullness and distension. It also increases appetite, reduces vomiting, diarrhea, cough, and phlegm.

Main Ingredients: The peel contains volatile oil, the major constituents of which is D-limonene, Citra, and α-pinene, etc.

11. 益智仁

Alpinia oxyphylla Miq.

科属：姜科，山姜属。

药名：益智仁。

别名：益智、益智子。

植物特征：多年生草本。地下茎丛生，地上茎直立，根茎短。叶2裂，互生，披针形或狭披针形，顶端尾尖，基部阔楔形，叶缘平直，具脱落性小刚毛。叶柄短，叶舌膜质，2裂，长1~2 cm。总状花序顶生，花蕾全包于一鞘状苞片中。雄蕊1枚，子房下位，卵圆形，3室。蒴果椭圆形或纺锤形，直径约1 cm。种子不规则扁圆形，棕黑色，被淡棕色假种皮。花期3~5月，果期4~9月。

资源状况：生长于林下阴湿处。主产于海南省，雷州半岛也有少量分布。广西、云南、福建等省（区）均有栽培种植。

药用部位：干燥成熟果实。

功能主治：有温脾、暖肾、固气、涩精之功效。主治寒性胃痛、脾虚吐泻、遗尿、尿频、遗精等症。

主要成分：果实含挥发油，益智酮A、B等。

Genus: Zingiberaceae, Alpinia.

Scientific Name: *Alpinia oxyphylla* Miq..

English Name: Sharpleaf galangal fruit.

Features: Perennial herbs. The underground stem is overgrown, the ground stem is erect and the root is short. Leaves have 2 columns, alternate, lanceolate or narrow lanceolate shape. The tail of apex is sharp, the root is wide cuneiform, leaf margin flatness, abscisic bristles. Short petiole, hyoid membrane, 2 cracks, 1-2 cm. Total inflorescence apex with buds all wrapped in a scabbard. 1 stamens, inferior ovary, ovoid shape, 3 ventricles. Capsule ellipsoid or spindle, about 1 cm. The seeds are irregular and round, brown and black, and are skinned in lightly browned. The florescence is from March to May. The fruit period is from April to September.

Resources Situation: It grows in the damp places of the forests. It is mainly produced in Hainan island, and Leizhou penin sula also has small amount of distribution. Guangxi, Yunnan, Fujian provinces (region) also cultivate it.

Medicinal Part: Dried ripe fruit or seeds.

Functions: Warming the spleen, warming the kidney and it has the effect of solid Qi, controlling nocturnal emission with astringent drugs. It is used to cure cold sex stomach pains, spleen deficient, vomiting and diarrhoea, enuresis, urinary frequency, spermatorrhea, etc.

Main Ingredients: The fruits contain Volatile oil, Alpinia oxyphylla ketone A, B, etc.

12. 降 香

Dalbergia odorifera T. Chen

科属：豆科，黄檀属。

药名：降香。

别名：降香黄檀、花梨木、紫藤香。

植物特征：乔木，小枝有密集白色的小皮孔。单数羽状复叶互生，小叶9~13片，近革质，卵形或椭圆形，顶端急尖，基部圆或宽楔形，托叶早落。圆锥花序腋生，花冠淡黄色或乳白色。旗瓣倒心状长圆形，雄蕊9枚，单体。荚果舌状长椭圆形，通常有种子1颗。

资源状况：原产于海南，是名贵的黄花梨木材和中药。广东有栽培，分布于粤东及粤西地区，属国家三级濒危保护植物。

药用部位：树干和根部心材。

功能主治：行气止痛，活血止血。可用于治疗吐血、咯血、风湿腰腿痛、痈疽疮肿、心胃气痛等。属理气药。

主要成分：主含挥发油、黄酮类等。

Genus: Leguminosae, Dalbergia.

Scientific Name: *Dalbergia odorifera* T. Chen.

English Name: Scented rosewood, Rosewood.

Features: This is a tree with white lenticels on small branches. It features odd pinnate leaves which alternate. It also has 9-13 oval leather-like leaflets, with an acute top and round or wedge-shaped base. The stipule falls early; axillary panicles, the corolla is faint yellow or ivory; the vexil is long and circular in shape, with nine monosome stamens. The ligulated legume is long elliptic usually with one seed inside.

Resources Situation: It is native to Hainan. It's a traditional Chinese herb and the tree of the rare Chrysanthemum pear wood. It's cultivated in Guangdong Province, is mostly distributed in the east and the west of Guangdong. It belongs to the third-class national endangered and protected plant species.

Medicinal Part: The Heartwood is considered the most medicinal part.

Functions: Promoting Qi circulation, relieving pain, promoting blood circulation, and stopping bleeding. It can be used to treat several symptoms such as spitting blood, hemoptysis, lumbocrural pain due to reumatism, ulcer and furuncle complications, heart-pains and stomachache.

Main Ingredients: Volatile oil, flavonoid.

13. 槟 榔

***Areca catechu* Linn.**

科属：棕榈科，槟榔属。

药名：槟榔。

别名：榔玉、宾门、青仔。

植物特征：常绿乔木，茎干直立，高10~18 m。有明显的环状叶痕，叶簇生于茎顶，长1.3~2 m，羽状复叶，叶片两面无毛，狭长披针形。雌雄同株，花序多分枝，雌花单生于分枝的基部，较大，花瓣近圆形；雄花小，无梗，通常单生，退化雄蕊6枚，合生。子房长圆形，果实长圆形或卵球形、橙黄色。种子卵形，花果期3~4月。

资源状况：原产于马来西亚，中国主要种植分布在云南、海南及台湾等热带地区。

药用部位：干燥成熟种子。

功能主治：杀虫消积，降气，行水，截疟。可用于治疗绦虫、蛔虫、姜片虫病，虫积腹痛，积滞泻痢，里急后重，水肿脚气，疟疾等。

主要成分：含生物碱、鞣质、脂肪及槟榔红色素。

Genus: Palmae, Areca.

Scientific Name: *Areca catechu* Linn..

English Name: Areca seed.

Features: Evergreen tree, up to 8-10 m. There are obvious annular leaf scars, and the foliage was born in stem top. The leaves are 1.3-2 m, most of them are pinnae, and both surfaces are glabrous. The shape of the leaves is narrowly lanceolate. It belongs to monoecious species and has branched inflorescence. The female flowers are born with basal part of the branches. The flowers are large and the shape of the patel are almost circle. The male flowers are small, sessile, and usually single. There are 6 stamens left when the flowers are degenerated. The stamens are connate.

Resources Situation: It is native to Malaysia. It is mainly distributed in the tropics of China, such as Yunnan, Hainan and Taiwan, etc.

Medicinal Part: The heartwood is considered the most medicinal part.

Functions: Promoting Qi circulation, relieving pain, promoting blood circulation, and stopping bleeding. It can be used to treat several symptoms such as spitting blood; hemoptysis; lumbocrural pain due to reumatism; ulcer and furuncle complications; heart-pains and stomachache.

Main Ingredients: Volatile oil, flavonoid.

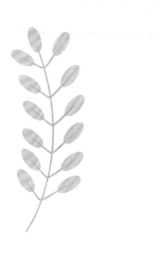

第三章
岭南主产中草药资源

除特有药材外，大部分的野生家种药材也是岭南主产药材。主要有银杏、杜仲、黄檗、厚朴、吴茱萸、蔓荆子、山药、泽泻、天花粉、使君子、穿心莲、紫苏、广东金钱草、鸡骨草、薄荷、水半夏、干姜、姜黄、山柰、壳砂仁、黄精、龙脷叶、芦荟、茯苓、灵芝等。

Chapter 3　Lingnan major Chinese herbal medicine resources

In addition to the unique medicinal materials, most of the wild domestic medicinal materials are also belong to Lingnan main medicinal materials. It mainly includes *Gingkgo Biloba*, *Eucommia ulmoides*, *Phellodendron amurense*, *Magnolia officinalis*, *Evodia ruticarpa*, *Vitex trifolia*, *Dioscorea polystachya*, *Alisma plantago-aquatica*, *Radices trichosanthis*, *Quispualis indica*, *Andrographis paniculata*, *Perilla frutescens*, *Desmodium styracifolium*, Lythrum salicaria, *Mentha canadensis*, *Typhonium cuspidatum*, *Zingiber officinale*, *Curcuma longa*, *Kaemmpferia galanga*, *Amomum villosum*, *Polygonatum sibiricum*, *Sauropus spatulifolius*, *Aloe vera var. chinensis*, *Wolfiporia extensa*, Ganoderma *lucidum*, etc.

1. 广东金钱草

Desmodium styracifolium (Osbeck) Merr.

科属：豆科，山蚂蟥属。

药名：广东金钱草。

别名：落地金钱、铜钱草、马蹄香。

植物特征：亚灌木状草本。茎直立或平卧，密被黄色柔毛。叶互生，披针形，托叶1对，小叶1~3片，圆形或长圆形，下面密被灰白色丝毛。顶生或腋生总状花序，苞片密集，每个苞片内有花2朵；花冠蝶形、紫红色、有香气。荚果被短柔毛和钩状毛，有3~6个荚节，每节有肾形种子1粒。花果期6~9月。

资源状况：分布于广东、广西、云南、贵州、四川以及越南、印度、缅甸。广东粤东及粤西地区均有栽培。

药用部位：全草。

功能主治：清热祛湿，利尿通淋。可用于治疗尿路感染、泌尿系结石、肾炎水肿等。

主要成分：全草含生物碱、酚类、鞣质。

Genus: Leguminosae, Desmodium.

Scientific Name: *Desmodium styracifolium* (Osbeck.) Merr..

English Name: Snowbell-leaf tickclover herb.

Features: Subshrublike herb. It has procumbent or erect stems which are thickly covered with yellow pubescence and alternating leaves, one pair of lanceolate stipule, 1-3 round or oblong leaflets tomentose on the lower surface, basidixed or axillary raceme, and dense bracts. There are 2 flowers in each bract. The papilonaceous corolla is prunosus with fragrance. The legume is covered by short pubescence barbs. Each pod contains 3-6 sections; each section contains one renal shaped seed. The florescence and fruit period is from June to September.

Resources Situation: *Desmodium styracifolium* (Osbeck.) Merr. is mostly distributed in Guangdong Province, Guangxi Zhuang Autonomous Region, Yunnan Province, Guizhou Province, Sichuan Province, Vietnam, India, and Myanmar. It has been planted in the eastern and western regions of Guangdong province.

Medicinal Part: The whole herb.

Functions: Clearing away heat and expelling dampness, relieving stranguria by diuresis. It is used to cure urinary tract infection, urinary system stone, nephritic edema, etc.

Main Ingredients: The whole plant contains alkaloids, phenols and tannin.

2. 蔓荆子

***Vitex trifolia* Linn.**

科属：马鞭草科，牡荆属。

药名：蔓荆子。

别名：蔓荆实、荆子、蔓青子。

植物特征：落叶灌木。叶对生，倒卵形，常为3出复叶，全缘。圆锥花序顶生，花序梗密被灰白色绒毛。花萼钟形，顶端5齿裂，花冠淡紫色或蓝紫色，顶端5裂，二唇形。果实近球形，成熟时黑色。花期7月，果期9~11月。

资源状况：单叶蔓荆子以惠阳、湛江、海南岛较多，生长于海边沙滩上。蔓荆子（灌木、3小叶复叶）则多分布于丘陵台地，四会、阳春等地均有栽培。

药用部位：果实。

功能主治：疏散风热，清利头目。可用于治疗风热感冒、偏头痛、齿痛、赤眼、目睛内痛等。属解表药下属分类的辛凉解表药。

主要成分：果实含蔓荆碱、卫矛醇、香草酸等。

Genus: Verbenaceae, vitex.

Scientific Name: *Vitex trifolia* Linn..

English Name: Fructus viticis.

Features: The plant is a shrub with opposing ternately compound obovate leaves, within an entire margin. The peduncle is densely covered by offwhite tomentum and acrogenous panical. It has a bell-like calyx, with 5-lobed tops. The corolla is light-purple or bluish-violet with 2 lips, and 5 cracks on the top. The subglobose drupe is black when it's ripe. The florescence is in July. The fruit period is from September to November.

Resources Situation: Simple fructus viticis mainly grows on seaside beaches in Huiyang, Zhanjiang and Hainan Island. *Vitex trifolia* Linn. (shrub or three compound leaves) is mostly distributed along the hills and plateaus in Sihui and Yangchun.

Medicinal Part: The fruit is considered the most medicinal part of the plant.

Functions: Dispelling wind and heat, clearing and disinhibiting head and eye pressure. It is used to cure antipyretic cold, headaches, migraines, toothache, red eyes and mesocalgia. It is one of Chinese herb medicines with the cool pharmaceutical hature.

Main Ingredients: The fruit contains vitex trifolia alkali, dulcite and vanillic acid, etc.

3. 穿心莲

***Andrographis paniculata* (Burm.f.) Nees**

科属：爵床科，穿心莲属。

药名：穿心莲。

别名：一见喜、苦胆草、四方莲。

植物特征：一年生草本。茎四棱形。叶卵状、至矩圆状披针形。总状花序顶生和腋生，花萼裂片三角状披针形。花冠白色而下唇带紫色斑纹，花冠筒与唇瓣等长。蒴果扁，成熟后开裂为2果瓣，种子12颗，多皱纹。花期8~9月，果期10月。

资源状况：主要栽培于广东、广西、福建等省区。现云南、四川、江西、江苏等省也有栽培。

药用部位：全草。

功能主治：清热解毒，凉血消肿。可用于治疗急性菌痢、胃肠炎、感冒、流脑、气管炎、口咽肿痛、疮疖痈肿等。

主要成分：二萜内酯类化合物和黄酮。

Genus: Acanthaceae, Andrographis.

Scientific Name: *Andrographis paniculata* (Burm. f.) Nees.

English Name: Common androgrphis herb, Herba andrographitis.

Features: It is an annual herb with 4 prismatic stems. The leaves are oblong or oblong lanceolate. It has a basidixed or axillary raceme. The calyx lobes are triangular lanceolate. The corolla is white and the labium has purple stripes. The corolla tube is as long as the labellum. The capsules are flat, it dehisces to 2 petals after maturity, and 12 wrinkled seeds. The flovescence is from August to September, and the fruit period is in October.

Resources Situation: *Andrographis paniculata* (Burm. f.) Nees is mainly planted in Guangdong, Guangxi, Fujian, etc. Now it's also been successfully transplanted in Yunnan, Sichuan, Jiangxi, and Jiangsu provinces.

Medicinal Part: The whole herb.

Functions: Detoxification. Lowering blood pressure, and reduceing swelling. It is used to cure acute bacillary dysentery, gastroenteritis, cold, epidemic cerebrospinal meningitis, trachitis, sore pharynx oralis, furuncle infections, and sores, etc.

Main Ingredients: Diterpene lactone and flavone.

4. 溪黄草

***Isodon Serra* (Maxim.) Kudo**

科属：唇形科，香茶菜属。

药名：溪黄草。

植物特征：多年生纤弱草本，下部常匍匐生根。茎方柱形，叶对生，纸质，卵圆形或宽卵形。揉之有黄色液汁，花白色或粉红色，两性或杂性，排成顶生的圆锥花序，花萼钟状、二唇形。小坚果卵状长椭圆形。秋季开花。

资源状况：线纹香茶菜为中药溪黄草的主要基原植物，生长主要分布于长江以南地区，广东的粤东、粤西地区均有栽培。

药用部位：干燥全草。

功能主治：清热利湿、退黄、凉血散瘀。可用于治疗湿热黄疸、湿热泻痢、跌打瘀肿。属清热药下属分类的清热燥湿药。

主要成分：含黄酮苷、酚类、氨基酸、有机酸等。

Genus: Labiatae, Rabdosia.

Scientific Name: *Isodon serra* (Maxim.) Kudo.

English Name: Linearstripe rabdosia herb.

Features: Perennial delicate herb with stoloniferous rhizome. The stems are rectangular columnar. The leaves are opposite, papery, orbicular-ovate or broad-ovate. The leaves excrete yellow juice when handled. The flowers are white or pink, bisexual or polygamous, and arranged into an acrogenous panicle. The calyx is bell-like and bilabiate. The small nut is egg-shaped or long oval. The plant blooms are in autumn.

Resources Situation: *Isodon serra* (Maxim.) Kudo is mostly from *Isodon lophanthoides* (Buch.-Ham.ex D.Don) H.Hara. It is mostly distributed in the southern area of the Changjiang River. It is planted in the eastern and western regions of Guangdong Province.

Medicinal Part: The whole dried herb.

Functions: Regulating body temperature and promoting diuresis. It also removes jaundice and promotes homeostasis. It's used to cure jaundice due to damp-heat, diarrhea, bruises and stasis swelling. It belongs to hat clearing and damp-drying drug.

Main Ingredients: Flavone glycosides, phenols, amino acid, and organic acids, etc.

5. 龙脷叶

Sauropus spatulifolius Beille

科属：大戟科，守宫木属。

药名：龙脷叶。

别名：龙舌叶、龙味叶。

植物特征：常绿小灌木。单叶互生，常聚生于小枝顶端，托叶三角形。叶片倒卵状长圆形或倒卵状披针形，基部狭窄。花细小，暗紫色，簇生于叶腋内。雌雄同枝，花萼6裂、二轮，近等大，倒卵形。蒴果扁球形，外有宿萼包被。花期2~10月。

资源状况：多栽培于药圃、公园、村边及屋旁。主要栽培分布于福建、广东、广西等省区。广东省主产地是肇庆高要。

药用部位：叶。

功能主治：清热化痰、润肺止咳、通便。可用于治疗肺燥咳嗽、咯血、大便秘结。属化痰止咳平喘药下分类的止咳平喘药。

Genus: Euphorbiaceae, Sauropus.

Scientific Name: *Sauropus spatulifolius* Beille.

English Name: Dragon's tongue leaf, Rostrate sauropus leaf.

Features: Evergreen undershrub. The plants have single alternating leaves, often crowded on the top of branches, with triangle stipules. The leaves are obovate long circles or obovate lanceolate with a narrow base. Tiny, dark-purple flowers cluster at the leaf axils. It has bisexual branch. There are 6 cracks in the calyx which has two rings. It is obovate. The fruits are oblate spheroid and have external calyx envelope. The florescence is from February to October.

Resources Situation: *Sauropus spatulifolius* Beille is planted in medicinal gardens, parks, villages or near the side of residences. It is distributed in Fujian Province, Guangdong Province, and Guangxi Zhuang Autonomous Region. The main producing area in Guangdong Province is Gaoyao, Zhaoqing.

Medicinal Part: The leaves are the most medicinal part of the plant.

Functions: Sweet, tasteless, or plain. It is known for eliminating phlegm by detoxification, stopping cough, nourishing lungs and relaxing the bowels. It's used to cure lung dryness and cough, hemoptysis and constipation.

6. 山 奈

***Kaempferia galanga* Linn.**

科属：姜科，山奈属。

药名：山奈。

别名：砂姜、三奈。

植物特征：多年生草本，根状茎块状，淡绿色或绿白色、芳香。叶通常2片贴近地面生长，近圆形。穗状花序4~12朵，顶生，蒴果。花期8~9月。

资源状况：海南、广西、云南、台湾等均有分布种植。主产于广东，粤西、粤东均有栽培。

药用部位：根茎。

功能主治：温中止痛、行气消食。可用于治疗胸膈胀满、脘腹冷痛、食积不化。外用可用于治疗跌打损伤等。

主要成分：含挥发油，油中主要成分为桂皮酸乙酯等，也含黄酮类成分山奈酚、山奈素等。

Genus: Zingiberaceae, Kaempferia.

Scientific Name: *Kaempferia galanga* Linn..

English Name: Galanga resurrectionlily.

Features: Perennial herb with fragrant tuberous rootstock. Usually two leaves grow close to the ground. The leaves are suborbicular with 4-12 spicate acrogenous flowers. The fruit is encapsulated with light green and white color. The florescence is from August to September.

Resources Situation: It is distributed in Hainan, Guangxi, Yunnan, and Taiwan. It is mainly produced in Guangdong Province. It's planted in the western and eastern regions of Guangdong.

Medicinal Part: Rootstock.

Functions: Warming the spleen and stopping pain, promoting blood circulation and reducing bloating. It also increases metabolism. It's used to treat chest and diaphragm pressure, crymodynia in stomach duct and abdomen, and indigestion. It also has external effects for traumatic injuries.

Main Ingredients: Volatile oil which contains ethyl cinnamate, flavonoids like kaempferol and kaempferide, etc.

7. 蔓性千斤拔

Flemingia philippinensis Merr.et Rolfe

科属：豆科，千斤拔属。

药名：蔓性千斤拔。

别名：老鼠尾、牛大力、独角龙、土黄鸡、金鸡落地。

植物特征：多年生直立或平卧亚灌木。幼枝三棱柱状，密被紧贴丝质柔毛，叶具指状3小叶，托叶线状披针形，有纵纹，被毛，先端细尖，宿存；小叶厚纸质，长椭圆形或卵状披针形，上面被疏短柔毛，背面密被灰褐色柔毛；基出脉3，侧脉及网脉在上面凹陷，下面凸起，侧生小叶略小；小叶柄极短，密被短柔毛。总状花序腋生，各部密被灰褐色至灰白色柔毛；苞片狭卵状披针形；花密生，具短梗；萼裂片披针形，远较萼管长，被灰白色长伏毛；花冠紫红色，荚果椭圆形，被短柔毛；种子近圆球形，黑色。花期、果期夏秋季。

资源状况：蔓性千斤拔主产于云南、贵州、四川、江西、福建、台湾、广东、海南、广西等省区。印度、孟加拉、缅甸、老挝、越南、柬埔寨、马来西亚、印度尼西亚亦有种植。常生长于旷野草地上或灌木丛中，山谷路旁和疏林阳处也可生长。

药用部位：根。

功能主治：祛风活血、强腰壮骨，可用于治疗风湿骨痛。以蔓性千斤拔的根作为原料生产的中成药有妇科千金片、金鸡冲剂、壮腰健肾丸等。

主要成分：以染料木素为代表的异黄酮类化合物。

Genus: Leguminosae, Flemingia.

Scientific Name: *Flemingia philippinensis* Merr.et Rolfe.

English Name: Philippine Flemingia root.

Features: Perennial erect shrub. The shape of the branch is like triangular prism, densely adhered to silky soft hair. The shape of the leaves is like 3 small fingers. The stipules are linear lanceolate, with longitudinal lines, hairs, and cuspidal front end. The leaflets are like thick paper, oval or ovate-lanceolate. The above is sparsely pubescent hair, and the back is densely covered with gray brown pubescence. The basal veins are 3. The upper side of the the side veins and the net veins are cavernous and the other side is protruding. The lateral leaflets are slightly small, the petiole is very short, and covered with densely pubescence. The racemes are axillary and densely covered with taupe and hoar pubescence. The bracts are narrowly ovate and lanceolate. It has dense flowers and short stalks. Calyx lobes are lanceolate, it has long calyx tube, covered with hoar adpressed hair. The corollas are purple and red, the pods are oval, covered with short pubescence. The seeds are nearly round and the color is black. It blooms in the summer and beats fruit in autumn.

Resources Situation: *Flemingia philippinensis* Merr.et Rolfe is mostly produced in Yunnan, Guizhou, Sichuan, Jiangxi, Fujian, Taiwan, Guangdong, Hainan, Guangxi, etc. It is also planted in India, Bangladesh, Burma, Laos, Vietnam, Cambodia, Malaysia and Indonesia. It often grows on the grass or bushes of the desert, valley along the road and sparse forests with sun.

Medicinal Part: Root.

Functions: Dispelling wind and circulating the blood, strengthing the waist and bone, preventing rheumatism bone pain. Patent medicines with flemingia macrophylla as the raw material: Gynecologic Qianjin tablets, Jinji electuary, Zhuangyao Jianshen pills.

Main Ingredients: Isoflavones like Genistein.

8. 岗梅根

***Ilex asprella* (Hook. et Arn.) Champ. ex Beth.**

科属：冬青科，冬青属。

药名：岗梅根。

别名：苦梅根、秤星树、假青梅。

植物特征：落叶灌木，高达3 m。枝条表面散生多数白色皮孔。叶互生、叶片卵形或卵状椭圆形，边缘具细锯齿，花白色，雌雄异株。雄花2~3朵簇生或单生于叶腋，萼片和花瓣各4~5片；雌花单生于叶腋，花瓣通常4片，基部合生。果为浆果状核果，圆球形，成熟时为黑色。内有分核4~6粒，内果皮骨质。花期3月，果期4~10月。

资源状况：岗梅根主产于广东、广西、福建、台湾、江西、湖南等省区；菲律宾吕宋、琉球等地亦有分布种植。生长于山坡草丛、路旁及次生林绿野径旁等环境，多生长在海拔400～1 000 m的地区。性喜高温，全日照或半日照均可，以腐殖质土壤生长最佳，多以种子繁殖。

药用部位：根。

功能主治：清热解毒，消肿止痛，具有抗菌作用。可用于治疗感冒、肺脓肿、急性扁桃体炎、咽喉炎、颈淋巴结结核、跌打损伤等。

主要成分：含三萜皂苷、内脂等。

Genus: Aquifoliaceae, Ilex.

Scientific Name: *Ilex asprella* (Hook. et Arn.) Champ. ex Beth..

English Name: Roughhaired holly root.

Features: Deciduous shrubs and up to 3 m. The surface of the branch is scattered with numerous white skin holes. Alternate leaves, leaf ovate or ovate-elliptic, serrate margin, white, dioecious strain. Male flowers are with 2-3 clusters or single leaf axillary, sepal and petal each has 4-5 tablets. The female flower is born in the axillary, usually there are 4 petals, and the root is in the same place. Fruit is a berry fruit, spheroidal and ripe, black. There are 4-5 kernels of endocarp and bone in the endocarp. The florescence is in March, and the fruit period is from April to October.

Resources Situation: *Ilex asprella Ilex asprella* (Hook. et Arn.) Champ. ex Beth. is mostly produced in Guangdong, Guangxi, Fujian, Taiwan, Jiangxi, Hunan, etc. Luzon in the Philippines and Ryukyu islands and other places also have distribution. It is growing in the grass slope, road and secondary wizard environment and places with altitude of 400-1,000 m. Like high temperature, full or half sunshine, grow best in humus soil and mostly with seed breeding.

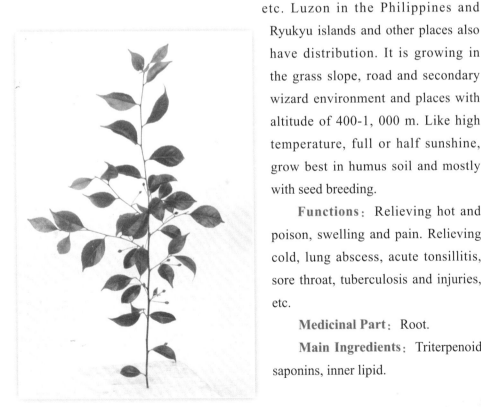

Functions: Relieving hot and poison, swelling and pain. Relieving cold, lung abscess, acute tonsillitis, sore throat, tuberculosis and injuries, etc.

Medicinal Part: Root.

Main Ingredients: Triterpenoid saponins, inner lipid.

9. 凉粉草

Mesona chinensis Benth.

科属：唇形科，凉粉草属。

药名：凉粉草。

别名：仙人草、仙人冻、仙草。

植物特征：草本。茎下部伏地，上部直立，茎、枝四棱形。叶狭卵圆形或近圆形，边缘有小锯齿，两面均有疏长毛。花冠白色或淡红色，内面在上唇片下方冠筒内略被微柔毛，冠檐二唇形，上唇具4齿，2侧齿较高，中央2齿不明显，下唇全缘，舟状。小坚果长圆形，黑色。花、果期7~10月。

资源状况：凉粉草主产于台湾、浙江、江西、广东及广西西部；生于水沟边及干沙地草丛中。

药用部位：干燥全草。

功能主治：消暑，解渴，除热毒。可用于治疗中暑，消渴，高血压，肌肉、关节疼痛等。

主要成分：含凉粉草多糖。

Genus: Labiatae, Mesona.

Scientific Name: *Mesona chinensis* Benth..

English Name: Herb of chinese mesona.

Features: It is herb. The lower part of the stems is prostrate on the ground, the upper part is erect. Stems and branches are prismatic shape. Leaves blade narrowly ovate to subcircular, margin with small serrate, both sides have thinning hairs. White or reddish corolla , the inner surface is slightly pubescent under the lower crown of the lower lip. The limb is with a double lip shape, upper lip with 4 teeth, and high teeth on both sides.The 2 teeth in the central are no obvious, lower lip with full margin and boat shape. Nutlets are oblong, black. The florescence and fruit period is from July to October.

Resources Situation: *Mesona chinensis* Benth. is mostly produced in Taiwan, Zhejiang, Jiangxi, Guangdong and western Guangxi and born in the grass by the ditch and in the dry sand.

Medicinal Part: The whole dried herb.

Functions: Relieving heat, thirst and toxic heat. It can cure heat stroke, diabetes, high blood pressure, muscle and joint pain, etc.

Main Ingredients: Bean jelly herb polysaccharides.

10. 牛大力

***Millettia speciosa* Champ.**

科属：豆科，崖豆藤属。

药名：牛大力。

别名：猪脚笠、金钟根、山莲藕、倒吊金钟、大力薯。

植物特征：藤本，长1~3 m。根系向下直伸，长约1 m。幼枝有棱角，被褐色柔毛，渐变无毛。叶互生；奇数羽状复叶，叶柄长

3~4 cm，托叶披针形，宿存，小叶7~17片，具短柄，基部有针状托叶一对，宿存；叶片长椭圆形或长椭圆披针形，先端钝短尖，基部钝圆；上面无毛，光亮，干后粉绿色；下面被柔毛或无毛，干后红褐色，边缘反卷。花两性，腋生，短总状花序稠密；花苞2裂；萼5裂，披针形；花冠略长于萼，粉红色，旗瓣秃净，圆形，基部白色，外有纵紫纹；翼瓣基部白色，有柄，前端紫色；龙骨瓣2片，基部浅白色，前部互相包着雌雄蕊；雄蕊10，两体，花药黄色，圆形；雌蕊1，子房上位。荚果长8~10 mm，径约5 mm。种子2枚，圆形。花期7~10月。果期次年2月。

资源分布：牛大力生长在深山幽谷之中。在福建、湖南、广东、广西等地有栽培。

药用部位：干燥根。

功能主治：可用于治疗腰肌劳损，风湿性关节炎，肺热、肺虚咳嗽。

主要成分：根含多种黄酮类化合物，如蔓性千斤拔素、羽扇豆醇等。

Genus: Leguminosae, Millettia.

Scientific Name: *Millettia speciosa* Champ..

Features: Climbing shrubs, 1-3 m length. The root is horizontal, about 1 m length. The young branchlets are angular, with brown pubescent and gradually glabrous. Alternate leaves and odd pinnate compound leaves, the petioles are 3-4 cm length. The stipule is lanceolate, persistent. There are 7-17 lobules with short stalks. A pair of needle shaped stipules are at the base, persistent. The leaves are oblong or oblong lanceolate. The leave tips are sharp. The base parts have blunt ends. The upper sides of the leaves are glabrous, bright, and become green after it is dry. The other sides of the leaves are glabrous or covered with pubescence. They become reddish brown after it is dry. The edges of the leaves are reflexed. The flowers are hermaphroditic, axillary, and the short racemes are dense. Buds crack for two parts. Calyxes crack for 5 parts, lanceolate. Corollas are pink and a little longer than calyxes. The vexils are bald and round. The base part is white with vertical purple strips. There are 2 keel petals with light white base. The pistils and stamens are wrapped in the front part. It has 10 diadelphous stamens. It has 1 pistil and superior ovary. The anthers are yellow and round. The pods are 8-10 m long, with a diameter of about 5 millimeters. There are 2 oval seeds. It blooms flowers from July to October and bears fruits in February of the following year.

Resources Situation: *Millettia speciosa* Champ. grows in the wet place near the stream of the forest and cultivated in Fujian, Hunan, Guangdong, Guangxi, etc.

Medicinal Part: Dried root.

Functions: Relieving sore throat, asthma, coughing, vomiting blood, furuncle swollen poison.

Main Ingredients: Including various flavonoids compounds, such as cranberry extract, lnpeol, etc.

11. 香 茅

Cymbopogon citratus **(DC.) Stapf**

科属：禾本科，香茅属。

药名：香茅。

别名：香茅草、大风茅。

植物特征：多年生草本。全株具柠檬香气，秆直立，粗壮，丛生，节常有蜡粉。叶片线形，无毛，边缘粗糙；叶鞘无毛；叶舌鳞片状。圆锥花序疏散，由成对、下托佛焰苞的总状花序所组成，分枝纤细；小穗成对，均无芒。花果期夏季，少见有开花者。

资源状况：原产于东南亚热带地区，喜高温多雨的气候，在无霜或少霜的地区都生长良好。广泛种植于热带地区，西印度群岛与非洲东部也有栽培。

药用部位：干燥全草。

功能主治：祛风解表，活血通络，行水消肿。用于治疗外感风寒头痛、胃寒痛、风湿痹痛等。

主要成分：主要含挥发油，油中主要成分为柠檬醛及香叶烯。

Genus: Gramineae, Cymbopogon.

Scientific Name: *Cymbopogon citratus* (DC.) Stapf.

English Name: Lemongrass herb.

Features: Perennial herbs. The whole plant has the aroma of lemon, the stem is erect, hairchested, tufted, the section often has wax powder. Leaf blade is linear shape, glabrous and rough margin. Leaf sheath is glabrous; ligule is scaly. Conic inflorescence evacuation, consisting of the total inflorescence of pairings and lower toffees, branched and slender. Small panicle is in pairs without awn. The florescence is in summer and few flowers bloom.

Resources Situation: It is native to Southeast Asia tropics, and is fond of high temperature and rainy weather. In places where without frost or less frost it grows much more better. It is widely planted in the tropical areas. It is also planted in the west Indies and east Africa.

Medicinal Part: The whole dried herb.

Functions: Dispelling wind and relieving exogenous syndrome, promoting blood circulation to remove meridian obstruction, diuresis and detumescence. It is used to cure cold exogenous, cold headache, stomach pain, rheumatic pain, etc.

Main Ingredients: Volatile oil, the main ingredients of oil are citral and myrcene.

12. 五指毛桃

Ficus hirta Vahl

科属：桑科，榕属。

药名：五指毛桃。

别名：粗叶榕、三龙爪。

植物特征：灌木或落叶小乔木，全株被黄褐色贴伏短硬毛，有乳汁。叶互生；叶片纸质，多型，长椭圆状披针形或狭广卵形，常具3~5深裂片，微波状锯齿或全缘，两面粗糙，基生脉3~7条；具叶柄。隐头花序，花序托对生于叶腋或已落叶枝的叶腋间，球形，幼时顶部有苞片形成的脐状突起，基部苞片卵状披针形，被柔毛；雄花、瘿花生于同一花序托内；雄花生于榕果内壁近顶部，花被片4，线状披针形，雄蕊2~3枚；瘿花花被片与雄花相似，花柱侧生；雌花生于另一花序托内，花被片4。瘦果椭圆球形。花期5~7月，果期8~10月。

资源状况：生长于山林、山谷以及村寨沟的灌木丛中。分布于福建、广东、海南、广西、贵州、云南等。其中以广东河源人工种植面积最广。

药用部位：根。

功能主治：具有健脾补肺、行气利湿、舒筋活络之功，常用于治疗脾虚浮肿、食少无力、肺痨咳嗽、盗汗、带下、产后无乳、月经不调、风湿痹痛、水肿等症。

主要成分：含氨基酸、糖类、甾体、香豆精等。

Genus: Moraceae, Ficus.

Scientific Name: *Ficus hirta* Vahl.

English Name: Radix fici hirtae.

Features: Shrubs or deciduous small trees, 2 or 3 pieces are covered with yellowish brown hair with milk. Alternate leaves, paper texture leaf blade, polytype, long elliptical lanceolate or narrow broad ovate, often with 3-5 deep lobes, microwave serrate or full margin, rough on both sides, with 3-7 veins. It has petiole. Hypanthium, inflorescence born in axillary or deciduous leaf axil, spherical, with bracts are formed like umbilical protrusion shape (at the beginning of its growth is particularly evident). The basal bracts are lanceolate oval shape, (appressed) pilose. The male, gall peanuts are in same inflorescence. Male flower is born near the top of the intine of the banyan fruit with 4 flowers. It is linear lanceolate shape with 2-3 stamens. The galls are similar to the male flowers and grow at the side of the flowers. The female flowers are born in another inflorescence, and have 4 flowers. The lean fruit is elliptical spherical. The florescence is from May to July. The fruit period is from August to October.

Resources Situation: It grows in the mountains or bushes and ditches beside the village. It is distributed in Fujian, Guangdong, Hainan, Guangxi, Guizhou, Yunnan, etc. And the artificial planting area is the largest in Heyuan of Guangdong.

Medicinal Part: Root.

Functions: It has the function of invigorating spleen and lung, circulating Qi and eliminating dampness, relaxing tendons and activating collaterals. It is used to cure spleen deficient, edema, pulmonary consumption with cough, night sweat, leucorrhea, postparturient agalactia syndrome, irregular menstruation, rheumatic arthralgia, edema, etc.

Main Ingredients: Amino acids, sugars, steroids, sweet bean essence, etc.

13. 黄皮核

Clausena lansium **(Lour.) Skeels**

科属：芸香科，黄皮属。

药名：黄皮核。

别名：黄弹、黄皮子、黄皮果。

植物特征：常绿小乔木。幼枝、花轴、叶轴、叶柄及嫩叶背脉上均有凸起细油且密被短直毛，有香味。奇数羽状复叶互生；小叶片5~13，顶端1枚最大，向下逐渐变小，卵形或卵状椭圆形，先端锐尖或短渐尖，基部宽楔形，两侧不对称，边浅波状或具浅钝齿。聚伞状圆锥花序顶生或腋生，花枝扩展，多花；萼片5，广卵形；花瓣5，白色，匙形，长不超过5毫米，开放时反展；雄蕊10，长短互间；子房上位，5室，密被直长毛。浆果球形或卵形，淡黄色至暗黄色，密被细毛；果肉乳白色，半透明；种子绿色。花期4~5月，果期7~8月。

资源状况：分布于西南地区及福建、台湾、广东、海南、广西等地。

药用部位：种子。

功能主治：疏风解表，除痰行气；散结，止痛，消食。可用于治疗温病发热、咳嗽痰喘、气胀腹痛、黄肿、疟疾等症。

主要成分：果含黄皮新肉桂酰胺A、B、C、D；种子含油。

Genus: Rutaceae, Clausena.

Scientific Name: *Clausena lansium* (Lour.) Skeels.

English Name: Wampee seed, Wampee fruit.

Features: Evergreen trees. Young branches, axles, leaf shafts, petioles, and the reverse veins of the leaves are dense short hairs with raised fine points with fragrance. Odd pinnate compound leaves; 5-13 small leaves, the apex has the largest one, descending gradually smaller, ovate or elliptical ovate, apex acute or short acuminate, base wide wedge, asymmetric at both sides, shallow wavy or dull teeth. Polyumbellate panicle inflorescence apex or axillary, flowering branchlets and flowers; 5 sepals, broad ovate; 5 petals, white, spoon shape, less than 5 mm, during the open time is anti-exhibition; 10 stamens, interjangdan; ovary upper, 5 units, densely straight long hairy. Berry spherical or ovate, pale yellow to dark yellow, densely thin hairy. The seeds are green. The florescence is from April to May, the fruit period is from July to August.

Resources Situation: It is distributed in the southwest area, Fujian, Taiwan, Guangdong, Hainan, Guangxi, etc.

Medicinal Part: Seed.

Functions: Expelling wind and resolving the exterior, eliminating phlegm and circulating Qi. Removing stasis, relieving pain and helping digestion. It is used to cure dampness and fever, cough and excessive phlegm asthma, inflatable abdominal pain, swelling, malaria, etc.

Main Ingredients: The fruit has lansiumamide A, B, C, D. The seed contains oil.

第四章
岭南引种进口中草药资源

岭南引种进口药材主要通过海南省、广东省南部等地区从东南亚等国家逐步引入我国。主要品种有檀香、丁香、玉桂、玉果、苏木、儿茶和胖大海等。

Chapter 4　Lingnan import Chinese herbal medicine resources

The Lingnan imported medicinal materials are mainly introduced through Hainan, southern Guangdong from southeast Asia and other countries. The main species are *santalum album*, *Ocimum gratissimum* var. *suave*, *cinnamo mum aromaticum*, *Myristica fragrans*, *Caesalpinia sappan*, *Acacia catechu*, *sterculia lychnophora*, etc.

1. 檀 香

***Santalum album* Linn.**

科属：檀香科，檀香属。

药名：檀香。

别名：白檀、白檀木。

植物特征：常绿乔木。半寄生。单叶对生，叶片膜质，椭圆状卵形，顶端锐尖，基部阔楔形，背面有白粉。三歧聚伞式圆锥花序腋生或顶生；花被钟形，先端4裂；核果球形，成熟时深紫红色至紫黑色。花期5~6月，果期7~9月。

资源状况：檀香主产于印度、印度尼西亚、马来西亚等地。广东省自1962年开始引种栽培，经近30年的试验研究，推广栽培，已经总结出一套成熟的栽培技术。原试种于高州、遂溪、电白、徐闻等地。近几年来有较大面积的发展，目前主要在阳西、德庆、高要、廉江、遂溪、化州、电白等市、县，广州及周边地区种植。

药用部位：干燥心材。

功能主治：理气，和胃，止痛。主治胸腹疼痛、气逆、呕吐、冠心病、胸中闷痛。

主要成分：含挥发油（白檀油）。

Genus: Santalaceae, Santalum.

Scientific Name: *Santalum album* Linn..

English Name: Sandalwood, White sandalwood.

Features: It is a Samiparasite evergreen tree with simple membrane leaves. The leaflets are oval with acute apexes and base with wedge. The leaflet has white power in the back. The Trichotomy thyrse is sequence altar or basifixed. The perianth is bell-like with 4 tiny cracks near the tip; the spherical drupe is dark purple to purple or black when it's ripe. The florescence is from May to June, the fruit period is from July to September.

Resources Situation: *Santalum album* Linn. is mainly produced in India, Indonesia, and Malaysia, etc. It has been introduced and cultivated in Guangdong since 1962. After 30 years of trial research and extending cultivation, people in Guangdong have concluded a set of mature cultivation techniques. In the beginning, it was only planted experimentally in Gaozhou, Suixi, Dianbai and Xuwen. In recent years, it has been planted in more areas. At present, it is mainly planted in Yangxi, Deqing, Gaoyao, Lianjiang, Suixi, Huazhou, Dianbai, Guangzhou, and the surrounding areas of Guangzhou.

Medicinal Part: Dry heartwood.

Functions: Promoting blood circulation, relieving stomach pain and bloating. It can also cure symptoms of chest pain, stomachache, reversed flow of Qi, vomiting, coronary heart disease, and oppressive chest pains.

Main Ingredients: Sandalwood oil.

2. 儿 茶

Acacia catechu **(L.f.) Willd.**

科属：豆科，金合欢属。

药名：儿茶。

别名：乌爹茶、孩儿茶、黑儿茶。

植物特征：落叶小乔木。嫩枝有刺，被短柔毛。托叶下常具一对钩状刺。二回羽状复叶，羽毛10~30对，叶轴被长柔毛；小叶20~50对，线形，被缘毛。穗状花序长2.5~10 cm，花淡黄色或白色；花萼钟状，萼齿三角形；花瓣披针形或倒披针形。荚果扁而薄，先端有喙，棕色，有光泽。花期4~8月，果期9月至翌年1月。

资源状况：原产于印度，在我国主要分布于浙江、台湾、广东、广西、云南。广东湛江南药试验场引种栽培。

药用部位：去皮枝、干的干燥煎膏。

功能主治：清热，生津，化痰，敛疮，止血，生肌。治痰热咳嗽，急性扁桃体发炎，湿疮，外伤出血，烧烫伤，水肿。（收湿生肌敛疮。用于溃疡不敛，湿疹，口疮，跌扑伤痛，外伤出血。）

主要成分：含儿茶鞣酸，儿茶精及表儿茶酚、黏液质、脂肪油、树胶及蜡等。

Genus: Leguminosae, Acacia.

Scientific Name: *Acacia catechu* (L.F.) willd..

English Name: Cutch.

Features: Deciduous small tree. Young branches are barbed and covered by short pubescence, and it has a pair of bipinnately compound leaves, and 10-30 pairs of accessory pinnae. The rachis is covered by pubescence; 20-50 pairs of linear leaflets which is covered by cilium. The Spica is 2.5-10 cm in length. The flowers are light yellow or white, bell-like shaped calyx, triangle calyx-tooth with lanceolate or oblanceolate petals. The legume is flat and thin, with a coronoid top which is brown and shiny. The florescence is from April to August. The fruit period is from September to the following year in January.

Resources Situation: It is native to India. It's mainly distributed in Zhejiang, Taiwan, Guangdong, Guangxi and Yunnan in China. It has been introduced and cultivated in the South China Medical Plants Testing Ground in Zhanjiang, Guangdong.

Medicinal Part: Peeled branches, dried and fried paste.

Functions: Clearing heat, promoting saliva secretion, reducing phlegm, heal abrasions and stops bleeding; it also promotes granulation. It's used to cure phlegm-heat syndrome, cough, acute tonsillitis, eczema, external bleeding, scalding burns, and edema.

Main Ingredients: It contains catechutannic acid, catechin, lepicatechol, phlegm, fatty oil, gum and wax, etc.

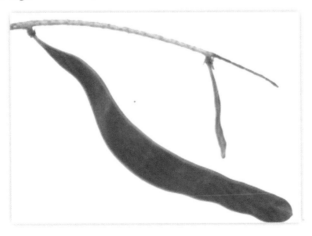

3. 大风子

***Hydnocarpus anthelminticus* Pierre ex Lanessan**

科属：大风子科，大风子属。

药名：大风子。

别名：大枫子、麻风子。

植物特征：常绿大乔木。单叶互生，卵状披针形或卵状长圆形，全缘或呈波状，无毛，幼叶紫红色。花杂性或单性；1至数朵簇生，花梗被短柔毛；萼片5，卵形，基部稍联合；花瓣5，卵形，黄绿色。浆果球形，果皮坚硬，有种子30~40粒。种子呈多角形或近卵形。花期9月，果期11月至次年6月。

资源状况：原产于越南、泰国及印度，湛江南药试验场引种栽培。

药用部位：种子。

功能主治：祛风燥湿，攻毒杀虫。治麻风病、癫疾、皮肤疥癣等。

主要成分：含多量脂肪油，油中主含大风子油酸、次大风子油酸和二者的甘油酯等。

Genus: Flacourtiaceae, Hydnocarpus.

Scientific Name: *Hydnocarpus anthelminticus* Pierre ex Lanessan.

English Name: Chaulmoogra, chaulmoogratree seed.

Features: It is an evergreen tree with simple leaves which alternates between a large or wavy margin. It is ovoid, lanceolate, or ovaloid. Young leaves are aubergine. Flowers are either polygamous or unisexual with clusters of one to several flowers. It is hairless. The pedicel is covered by short pubescence; 5 sepals, oval, somewhat united at the base. The 5 petals are yellowish-green ovals. The fruits are spherical berries with hard peel containing 30-40 seeds, which are polygonous or subovate. The florescence is in September, the fruit period is from November to the following year in June.

Resources Situation: *Hydnocarpus anthelminticus* Pierre ex Lanessan is native to Vietnam, Thailand and India. It has been introduced and cultivated in the South China Medical Plants Testing Ground in Zhangjiang, Guangdong.

Medicinal Part: Seeds.

Functions: Dispelling pathogens, removing excess water weigh, counteracting toxic substances, and killing worms. It's used to cure leprosy, favus of the scalp, and itchy skin.

Main Ingredients: It contains high fatty oil which consists of chaulmoogric acid, hydnocarpic acid and glyceride.

4. 马钱子

***Strychnos nux-vomica* Linn.**

科属：马钱科，马钱属。

药名：马钱子。

别名：番木鳖、苦实。

植物特征：落叶乔木。叶片纸质，对生，近圆形、宽椭圆形至卵形；聚伞花序顶生；花较小，灰白色；花冠筒状，先端5裂，裂片卵形。先端钝。浆果球形，成熟时橙色。种子1~4，圆盘形，表面灰黄色。密被银色绒毛。花期春夏两季，果期8月至次年1月。

资源状况：原产于印度，湛江南药试验场引种栽培。

药用部位：干燥成熟种子。大毒。孕妇禁用。运动员慎用，忌多服久服及生用。

功能主治：散结消肿，通络止痛。用于治疗风湿顽痹、咽喉痹痛、痈疽肿痛、麻木瘫痪等症。

主要成分：含番木鳖碱、马钱子碱、伪番木鳖碱等。

Genus: Loganiaceae, Strychnos.

Scientific Name: *Strychnos nux-vomica* Linn..

English Name: Quaker button, Vomic-nut semen. Strychni.

Features: Deciduous tree. The leaves are paper-like and opposite of each other. It is semi-round, wide elliptical to ovoid shape. The cyme is basifixed. The flowers are small and off-white. The corolla is tubular; on the obtuse tip, there are 5 breaches which are egg shaped. The berries are spherical.They turn orange when they are ripe. There are 1-4 seeds which are disc-shaped and grayish-yellow on the surface and densely covered by silver tomentum. The florescence is in spring and summer, and the fruit period is from August to the following year in January.

Resources Situation: It is native to India. It has been introduced and cultivated in the South China Medical Plants Testing Ground in Zhangjiang, Guangdong.

Medicinal Part: Dried and ripened seeds are the most medicinal part of this plant. It has strong toxicity. Pregnant women, athletes are forbidden to use. It can't take too much, take for a long time and take without cook.

Functions: Reducing swelling, activating meridians to alleviate pain. It's used to cure rheumatoid arthritis, sore throat, ulcers, numbness, and paralysis.

Main Ingredients: Strychnine, brucine, and Pseudostrychnine.

5. 诃 子

***Terminalia chebula* Retz.**

科属：使君子科，榄仁属。

药名：诃子。

别名：诃黎勒。

植物特征：落叶大乔木。单叶互生或近对生，叶柄粗壮；叶片椭圆形或卵形，两面近无毛。黄色花排成顶生的花穗状花序，花序轴被毛；花萼杯状，顶端5齿裂，内面有棕黄色长毛，无花瓣。核果椭圆形或近卵形，表面灰黄色或黄褐色，有5~6条钝裂。花期5月，果期7~9月。

资源状况：原产于东南亚各国，我国分布于广东、广西、云南等省区。湛江南药试验场引种栽培。

药用部位：成熟果实。

主要成分：果实含鞣质20%~40%。

功能主治：涩肠止血，敛肺止咳，清热利咽。用于治疗慢性咽喉炎、慢性肠炎、溃疡病、便血等症。

Genus：Combretaceae, Terminalia.

Scientific Name：*Terminalia chebula* Retz..

English Name：Fruetus chebulae.

Features：This plant is a deciduous tree with alternating or subopposite simple leaves. The petiole is rough; and the leaves are ovate or oval with both sides glabrescent.

The cyme is basified; the flowers are small and off-white. The yellow flowers from into the acrogenous spike. The inflorescence axle is covered with hairs on the obtuse tip, there are 5 breaches which are egg shape. The berries are spherical. They turn orange when they are ripe. There are 2-5 seeds which are disc-shaped and grayish-yellow on the surface and densely covered by silver fuzz. The florescence is in May, and the fruit period is from July to September.

Resources Situation：It is native to Southeast Asian countries, and is distributed in Guangdong, Guangxi, Yunnan and other provinces in China. It has been introduced and cultivated in the South China Medical Plants Testing Ground in Zhangjiang，Guangdong.

Medicinal Part：Ripened seeds are considered the most medicinal part of the plant.

Functions：Reducing swelling, activating meridians to stop pain. It's used to cure chronic pharyngolary rgits, rheumatoid arthritis, sore throat, ulcer pain, numbness and paralysis.

Main Ingredients：The fruit has 20%-40% tannin.

6. 苏 木

Caesalpinia sappan Linn.

科属：豆科，云实属。

药名：苏木。

别名：红柴、红苏木、苏枋、苏方木。

植物特征：灌木或小乔木蔓生，具刺，小枝微被柔毛，有皮孔。二回羽状复叶对生，羽片7~13对，叶轴被柔毛；小叶10~17对，长圆形，先端圆钝。圆锥花序顶生或腋生；花萼斜钟状，萼齿5，反卷；花冠黄色具红脉纹，花瓣5。荚果扁，木质，倒卵状长圆形。花期5~10月，果期7月至次年3月。

资源分布：原产于印度、缅甸、越南、马来西亚及斯里兰卡，在中国主要分布于华南地区。广东湛江南药试验场引种栽培。

药用部位：心材。

功能主治：行血，破瘀，消肿，止痛。用于治疗经闭腹痛、产后瘀血胀痛、跌打损伤等症。

主要成分：含挥发油。心材含巴西苏木素、苏木醇等。

Genus: Leguminosae, Caesalpinia.

Scientific Name: *Caesalpinia sappan* Linn..

English Name: Lignum sappan.

Features: Herb or small arbor with thorns. The plant sprawls with little sparsely lenticel puberulous twigs. Two pinnate leaves opposite, with 7-13 pairs, the rachis is puberulous, with 10-17 pairs of oblong leaflets, and an obtuse top. Panicels are basified or Axillary. The Calyx are oblique, bell-shaped, with 5 coiled outwards calyxtooth. The 5-petalled corolla is yellow with red veins. The legume is flat, woody, and obovate-oblong. The florescence is from May to October, and the fruit period is from July to the following year in March.

Resources Situation: It is native to India, Myanmar, Vietnam, Malaysia and Sri Lanka. It is mainly distributed in South China. It has been introduced and cultivated in the South China Medical Plants Testing Ground in Zhangjiang, Guangdong.

Medicinal Part: The heartwood is the most medicinal part of the plant.

Functions: Promoting blood circulation, reestablishing homeostasis, reducing swelling and pain relief. It is used to cure amenorrhea, stomachache, postpartum extraverted blood and swelling pain, and traumatic injuries.

Main Ingredients: This plant contains volatile oil. Heartwood contains Brazilian brasilin, and ethanol extract of Caesalpinia Sappan.

7. 胖大海

***Sterculia lychnophora* Hance**

科属：梧桐科，胖大海属。

药名：胖大海。

别名：大海子、大洞果、圆羊齿。

植物特征：落叶乔木。树皮粗糙而略具条纹。叶互生，叶片革质，卵形或椭圆状披针形，全缘，光滑无毛。花杂性同株，成顶生或腋生的圆锥花序；花萼钟状，宿存，裂片披针形；花瓣呈星状伸张；雄花具雄蕊10~15，罕至30，花药及花丝均被疏柔毛，不育心皮被短绒毛；雌花具1枚雌蕊。蓇葖果1~5个，着生于果梗，呈船形，在成熟之前裂开；最初被疏柔毛，后脱落。种子梭形或倒卵形，深黑褐色，表面具皱纹。

资源状况：生长于热带地区。主产于越南、泰国、印度尼西亚、马来西亚等国。我国云南西双版纳、海南有栽培。

药用部位：种子。

功能主治：清热，润肺，利咽，解毒。治干咳无痰、喉痛音哑、头痛、目赤、牙痛、痔疮等症。

主要成分：种子外层含西黄芪胶粘素，果皮含半乳糖15.06%，戊糖24.7%。

Genus: Sterculiaceae Scaphium.

Scientific Name: *Sterculia lychnophora* Hance.

English Name: Boat-fruited sterculia seed.

Features: Deciduous trees. The bark is rough and slightly striped. Leaves alternate, and it is leathery, ovate or elliptical lanceolate, whole margin, glossy glabrous. The conical inflorescence of flowering is basidixed or axillary. Calyx bell-shaped, persistent, lobes lanceolate; Petals extend like a star; Male flowers with stamens 10-15, and rarely up to 30. Anthers and silks are covered with pubic hair. Sterile core is short villi. The female fruit has 1 pistil. The fruit of follicle is 1-5. The fruit is born in peduncle, in the shape of a boat, opened before maturity; The first is the pubic hair, and then it falls off. Seed fusiform or obovate, dark brown, with wrinkles on the surface.The first is the pubic hair, and then it falls off.Seed fusiform or obovate, dark brown, with wrinkles on the surface.

Resources Situation: It grows in the tropical areas. It is mainly produced in Vietnam, Thailand, Indonesia, Malaysia, etc. It has been cultivated in Xishuangbanna, Yunnan and Hainan.

Medicinal Part: Seed.

Functions: Clearing heat, moistening lung, relieving sore throat and detoxify. It is used to cure dry cough without phlegm, sore throat, sound dumb, headache, hot eyes, toothache, hemorrhoids.

Main Ingredients: The seed contains outer west astragalus adhesive. The peel contains 15.06% galactose, 24.7% pentose.

第五章
岭南其他药用植物

岭南中药资源十分丰富，家养与野生资源品种众多。除特产、主产药材外，还包括许多珍稀濒危品种，以及常见待开发的植物，如有剧毒的见血封喉、具抗癌功效的喜树以及金毛狗、威灵仙等，具有巨大的开发空间。

Chapter 5　Lingnan other medicinal plants

Lingnan has abundant traditional Chinese medicine resources and various domestic and wild resources. In addition to the specialty and main medicinal materials, it also includes many rare and endangered species and common plant which needs to be developed, such as the *Antiaris txicaria* with highly toxic, *Camptotheca acuminata* with the function of anti-cancer, *Cibotium barometz*, Clematis chinensis, etc. There is huge space for development.

1. 见血封喉

***Antiaris toxicaria* Lesch.**

科属：桑科，见血封喉属。

药名：见血封喉。

别名：箭毒木。

植物特征：常绿大乔木，高可达40 m，通常具板状根；树皮灰色，略粗糙；小枝幼时被棕色柔毛，干后有皱纹。叶互生，二列，长圆形或长圆状椭圆形，全缘或具粗齿，表面亮绿色，疏生长粗毛，背面浅绿色，密被长粗毛，干后变为茶褐色。花单性，雌雄同株；雄花密集于叶腋，生于一肉质、盘状、有短柄的花序托上；雌花单生于具鳞片的梨形花序托内，无花被，子房与花序托合生，花柱2裂。果肉质，梨形，成熟时鲜红色至紫红色。花期3~4月，果期5~6月。

资源状况：该属有4种，生长在亚洲及非洲热带地区。我国只有见血封喉1种，分布于广东西部、海南、广西南部、云南西双版纳。属国家三级珍贵保护植物。

药用部位：乳汁或种子。

功能主治：强心苷中一些物质有强心、加速心律、增强心血输出作用，在医药上极有研究价值。近年来，医药专家研究其有效成分，用于治疗高血压，取得很好效果。

主要成分：见血封喉树皮和叶子具有白色乳汁，含强心苷（弩箭子苷、见血封喉苷、铃兰毒苷、铃兰毒醇苷、伊夫草苷、马来欧苷）等有毒物质。

Genus: Moraceae, Antiaris.

Scientific Name: *Antiaris toxicaria* Lesch..

English Name: Upas tree.

Features: Evergreen large tree with buttress roots. It can grow to a height of 40 m. The sprigs are covered by brown pubescene.It has wrinkle after it becomes dry. The leaves alternate in two columns. The leaves are long-circular or oval, and are rounded or serrated. The surface of the leaves are bright-green, and sparsely pilose-strigose while the reverse side is light green and densely covered by long thick tomentum. It is reverse tea brown after it becomes dry. The flowers are monogenic and monoecious. The male flowers are concentrated in leaf axils and born in a fleshy, discoid, short stalked inflorescence. Female flowers are born with a pear-shaped lepidote inflorescence, but without perianth. The ovary and receptacle are gambogenic. There are 2 gaps in the style. It produces pear-shaped fleshy fruit. The ripe fruit is bright red or aubergine. The florescence is from March to April. The fruit period is from May to June.

Resources Situation: This genus consists of 4 species which grows in tropical areas in Asia and Africa. Only *Antiaris toxicaria* Lesch. is found in China, and it is mainly distributed in western Guangdong, Hainan, southern Guangxi, and Xishuangbanna, Yunnan. It belongs to the third-class national rare and protected plant species.

Medicind Part: Juice or seed.

Functions: Certain substance in Cardiac Glycosides can be used to strengthen heart functions, accelerate heart rhythm, and enhance cardiac output. Many consider it to have great research potential. In the recent years, medical experts have been researching on the active ingredients in Cardiac Glycosides and have achieved great effects on treating hypertension.

Main Ingredients: The white milky juice in the bark and leaves of Antiaris toxicaria contain a poisonous substance like Cardiac Glycosides (antiarin, convalloside, perconval, Yves glycosides).

2. 喜 树

Camptotheca acuminata Decne.

科属：蓝果树科，喜树属。

药名：喜树。

别名：旱莲木、千丈树。

植物特征：落叶乔木。单叶互生，纸质，叶片长卵形或矩圆状椭圆形，顶端渐尖，基部宽楔形，全缘。花杂性，同株；淡绿色小花组成球形头状花序，再组成圆锥花序，顶生或腋生；花萼5浅裂，边缘有纤毛；花瓣5，外面密被短柔毛。果多数集成圆头形复果，核果窄长椭圆形，有窄翅。花期5~7月，果期9月。

资源状况：分布于我国南方各省，广东、广西有栽培。

药用部位：叶、果实、根及树皮。

功能主治：有毒。抗癌，散结，清热，杀虫。治慢性粒细胞白血病、胃癌、结肠癌、直肠癌等症。

主要成分：含抗肿瘤作用的生物碱，如喜树碱、10-羟基喜树碱等。

Genus: Nyssaceae, Camptotheca.

Scientific Name: *Camptotheca acuminata* Decne..

English Name: Dry lotus wood, Camptotheca.

Features: Deciduous tree. The leaves have long ovate shape, paper characteristics, with acuminate apexes, wedge-shaped bases, and with an entire spherical margin. The flowers are misceuaneous sex and monoecious. The small light-green flowers form a spherical inflorescence and then a circular cone shape, basifixed or axillary. There are 5 lobed places on the calyx and the edge is covered by cilia. All 5 petals, light green, are densely villous. The plant produces a round fruit and several commonly blossom together. Around the fruit are narrow long oval seeds with narrow wings. The florescence is from May to July, and the fruit period is in September.

Resources Situation: It is distributed in the provinces of Southern China and has been cultivated in Guangdong and Guangxi.

Medicinal Part: Root, leaf, fruit and bark.

Functions: Poisonous. Anticancer Removing stasis, clearing heat and pesticide ingredients. Furthermore, it can potentially treat chronic myelogenous leukemia, stomach cancer, colon cancer, and rectum cancer symptoms.

Main Ingredients: Contains antineoplastic alkaloid: camptothecin, 10-hydroxy camptothecin (hpt), etc.

3. 大叶骨碎补

Davallia formosana Hayata

科属：骨碎补科，骨碎补属。

药名：大叶骨碎补。

别名：硬骨碎补、华南骨碎补。

植物特征：多年生草本。根状茎粗壮横走，连同叶柄基部密生披针形、膜质、棕色鳞片。叶远生，纸质，无毛；叶柄及叶轴棕色；叶片三角形，4回或5回羽状分裂，顶部渐尖并为羽裂；羽片互生，有柄，基部一对最大，中部以上逐渐变小；小羽片有短柄，连同小羽轴有狭翅；末回裂片斜三角形，常二裂成不等长的尖齿。孢子囊群生于小脉中部稍下的弯弓处，或生于小脉分叉处；囊群盖盅形。

资源状况：附生于岩石或树干上。分布于广东、广西、台湾、云南。

药用部位：根茎。冬、春采挖，除去叶片及泥沙，晒干或蒸熟后晒干，用火烧去毛绒。

功能主治：补肾强骨，续伤止痛。用于治疗肾虚腰痛、耳鸣耳聋、牙齿松动、跌扑闪挫、筋骨折伤、外治斑秃、白癜风等症。

主要成分：含淀粉、葡萄糖、柚皮甙。

Genus: Davalliaceae, Davallia.

Scientific Name: *Davallia formosana* Hayata.

Features: Perennial herbs. The rhizome of the root is thick and strong, along with the base of the base of the petiole, the lanceolate, membranous, brown scales. Leaf near-born, paper, glabrous; Petiole and leaf axial brown; Leaf triangular, 4 or 5 back pinnate splitting, apex acuminate and feathered. The feathered slices alternate with each other, with a pair of stalks and a pair of base pairs. The small feather has a short shank with narrow wing; the end of the lobes oblique triangle, often with two different long sharp teeth. Sorus cysts is born in the lower vein in the middle and lower part of the vein, or in the small vein branch. The indusium is in cup shape.

Resources Situation: It grows on rocks or tree trunks. It is distributed in Guangdong, Guangxi, Taiwan and Yunnan.

Medicinal Part: Rhizome. Excavation in winter and spring, remove the blade and mud sand, dry or steamed dried, use fire to wipe off the hairy.

Functions: Strengthening kidney and improving bone, curing injuries and relieving pain. It is used to cure kidney deficiency and lumbar pain, tinnitus and deafness, gomphiasis, fall down and injury, broken bone and muscle, and external use for alopecia areata, vitiligo.

Main Ingredients: Starch, glucose, pomelo peel glucoside.

4. 狗 脊

Cibotium barometz (Linn.) J.Sm.

科属：蚌壳蕨科，金毛狗属。

药名：狗脊。

别名：金毛狮子、猴毛头、金毛狗脊、黄狗头。

植物特征：多年生树蕨。根茎平卧，有时转为直立，短而粗壮，带木质，密被棕黄色带有金色光泽的长柔毛。叶多数，丛生成冠状，大形；叶柄粗壮，褐色，基部密被金黄色长柔毛和黄色狭长披针形鳞片；叶片卵圆形，3回羽状分裂；亚革质，上面暗绿色，下面粉灰色，叶脉开放，不分枝。孢子囊群盖生于边缘的侧脉顶上，略成矩圆形，每裂片上2~12枚，囊群盖侧裂呈双唇状，棕褐色。

资源状况：生长于山脚沟边，或林下阴处酸性土壤。分布于我国南部、东南部、西南部及河南、湖北等地。主要产于四川、福建、浙江、广西、广东、贵州、江西、湖北等地。

药用部位：根茎、根茎上的长柔毛。

功能主治：补肝肾，除风湿，健腰脚，利关节。治腰背酸疼、膝痛脚弱、寒湿周痹、失溺、尿频、遗精、白带等症。

主要成分：含淀粉，含鞣质类。

Genus: Dicksoniaceae, Cibotium.

Scientific Name: *Cibotium barometz* (Linn.) J.Sm..

English Name: Rhizoma cibotii.

Features: A perennial tree fern. The roots lie flat, sometimes turn erect, short, thick, and wooden, densely covered with brown and yellow with a golden sheen of soft fur. Leaves with a lot of pieces, clumps into crowns, large; Petiole stout, brown, base densely covered with yellow long soft fur and yellow narrow lanceolate scales; ovoid leaf, 3 pinnate split; Subleathery, covered in dark green, gray with flour, open veins, unbranched. Sorus cysts is born on the edge of the lateral veins, and is slightly rounded with 2-12 pieces on each lobes, with a double lip and is brown.

Resources Situation: It grows in the foothills river bank, or forests in the shade of acidic soil. It is distributed in the south, southeast and southwest of China and Henan, Hubei, etc. It is mainly produced in Sichuan, Fujian, Zhejiang, Guangxi, Guangdong, Guizhou, Jiangxi, Hubei, etc.

Medicinal Part: Root, long hairs of the root.

Functions: Nourishing liver and kidney, eliminating rheumatism, enhancing waist and legs, lubricating joints. It is used to cure sore back, cold and dampness, arthralgia, urinary incontinence, frequent urination, nocturnal emission and leucorrhea.

Main Ingredients: Starchy, tannins.

5. 贯 众

Cyrtomium fortunei J. Sm.

科属：鳞毛蕨科，贯众属。

药名：贯众。

别名：神箭根、小晕头鸡、鸡脑壳、公鸡头。

植物特征：多年生草本。地下根茎斜生，粗人块状，坚硬，叶柄基部密被阔卵状披针形黑褐色大形鳞片。叶簇生，叶柄有疏鳞片；叶片长圆形至披针形，一回羽状；羽片有短柄，上侧稍呈尖耳状突起，边缘具细锯齿；叶脉网状。孢子囊群分布于叶片中部以上的羽片上，生于内藏小脉顶端，散生；囊群盖肾圆形，棕色。

资源状况：生长于林下沼地。分布于河北、江西、江苏、安徽、广东、广西、湖南等地。

药用部位：根状茎。

功能主治：清热解毒，凉血息风，散瘀止血，驱虫。治感冒、热病斑疹、吐血便血、乳痈、跌打损伤等。

主要成分：含异槲皮苷、紫云英苷、冷蕨苷、贯众苷、鞣质及挥发油等。

Genus: Dryopteridaceae, Cyrtomium.

Scientific Name: *Cyrtomium fortunei* J. Sm..

English Name: Arrow root, Faint chicken.

Features: Perennial herbs. The underground root is oblique, thick, hard, and the base of the petiole is densely ovate. Leaf clusters, petiole of scaly. The blade is round to the cape, and a feather. The pinna has a short handle. The upper side is slightly pointed, and the edge is serrated. A network of veins. The Sorus cysts is distributed in the pinna above the center of the leaf, and is born on the apex of the inner veins. The cysts are round and brown.

Resources Situation: It grows in marshland of forest and is distributed in Hebei, Jiangxi, Jiangsu, Anhui, Guangdong, Guangxi, Hunan, etc.

Medicinal Part: Rhizome.

Functions: Clearing heat and removing toxicity, cooling blood and calming endogenous wind, scattering stasis hemostasis and expelling parasite. It is used to cure cold, spotted fever, vomit blood and stool blood, acute mastitis and traumatic injury.

Main Ingredients: Isoquercitrin, astragalus sinicus glycosides, cold fern glycosides, cyrtomium rhizome glycosides, tannin and volatile oil, etc.

6. 威灵仙

Clematis chinensis Osbeck

科属：毛茛科，铁线莲属。

药名：威灵仙。

别名：百条根、老虎须、铁脚威灵仙、黑须公、白钱草。

植物特征：木质藤本。根多数丛生，细长，外皮黑褐。茎干后黑色，具明显条纹，幼时被白色细柔毛，老时脱落。叶对生，一回羽状复叶，小叶通常5片，罕为3片，小叶卵形或卵状披针形，全缘，上面沿叶脉有细毛，下面光滑，主脉3条。圆锥花序腋生及顶生；苞片叶状；萼片4或5，花瓣状，长圆状倒卵形，白色，顶端常有小尖头凸出，外侧被白色柔毛，内侧光滑无毛；雄蕊多数，不等长，花丝扁平；雌蕊4~6，心皮分离，子房及花柱上密生白色毛。瘦果扁平状卵形，略生细短毛，花柱宿存，延长呈白色羽毛状。花期6~9月，果期8~11月。

资源状况：生长于山野、田埂及路旁。产于江苏、安徽、浙江、山东、四川、广东、广西、湖南、福建等地。

药用部位：干燥根及根茎。

功能主治：祛风除湿，通络止痛。用于治疗风湿痹痛、肢体麻木、筋脉拘挛、屈伸不利、骨鲠咽喉等。

主要成分：根含白头翁素、白头翁内酯、甾醇、糖类、皂甙、内酯、酚类、氨基酸。叶含内脂、酚类、三萜、氨基酸、有机酸等。

Genus: Ranunculaceae, Clematis.

Scientific Name: *Clematis chinensis* Osbeck.

English Name: Aster turbinatus, Tacca chantrieri.

Features: Wooden vine. The roots are mostly thick, slender and black and brown. The stem is dried and black, with clear streaks, and when young, it is white and soft, and it falls off when it is old. Leaf pair, one feather compound leaf. Small leaf usually has 5 pieces, rarely with 3 pieces, lobular ovate or ovate-lanceolate, whole margin, upper along the leaf vein has fine hair, the bottom smooth, 3 main veins. Panicles and the apex of the axilla; 4 or 5 bract leaf sepals, petal, long rounded obovate, white, apex often with small pointed protruding, lateral white pubescent, inside smooth glabrous; Mostly are stamens, unequal length, filament flat; 4 pistils, heart skin separation, ovary and flower column densely white hairs. Thin fruit flattened ovate, slightly short hair, style of flowers, elongated white plumage. The florescence is from June to September, and the fruit period is from August to November.

Resources Situation: It grows in the mountains, ridges or roadsides. It is produced in Jiangsu, Anhui, Zhejiang, Shandong, Sichuan, Guangdong, Guangxi, Hunan, Fujian, etc.

Medicinal Part: Dried root and rhizome.

Functions: Dispelling wind and eliminating dampness, smoothing collaterals and relieving pain. It is used to cure rheumatic arthralgia, extremities numbness, spasm, inhibited bending and stretching, bone stuck in throat.

Main Ingredients:
The root contains anemonin, anemonol, sterol, saccharides, saponins, lactones, phenols, amino acids. The leaf contains lactone phenols, triterpene, amino acid, organic acid, etc.

7. 阴 香

Cinnamomum burmanni (Nees et T.Nees) Blume

科属：樟科，樟属。

药名：阴香。

别名：坎香草、阴草、山肉桂、假樟树。

植物特征：常绿乔木。小枝赤褐色，无毛。叶近于对生或互生，革质，卵形或长椭圆形，全缘，无毛，具离基3出脉。圆锥花序顶生或腋生；花小，绿白色；花被6，基部略合生，两面均被柔毛；能育雄蕊9，排成3轮，外面2轮花药内向，第3轮花药外向，花药均为卵形，4室，瓣裂，花丝短，最内尚有1轮退化雄蕊；雌蕊1，子房上位，1室，1胚珠。浆果核果状，卵形。花期8~11月，果期11月至次年2月。

资源状况：生长于疏林中有阳光处。分布于广东、广西、江西、浙江、福建等地。

药用部位：本植物的皮、根、叶。

功能主治：根，温中，散寒，祛风湿；治食少、腹胀、水泻、脘腹疼痛、风湿、疮肿、跌打扭伤等。叶，治风湿骨痛、寒湿泻痢、腹痛等。

主要成分：树皮含桂皮醛、丁香油酚、黄樟醚等挥发油；叶含丁香油酚、芳樟醇等挥发油。

Genus: Lauraceae, Cinnamomum.

Scientific Name: *Cinnamomum burmanni* (Nees et T.Nees) Blume.

English Name: Candy vanilla.

Features: Evergreen trees. The little twigs are auburn and hairless. Leaves close to opposite or alternate, leathery, ovate or long elliptic, full-margin, glabrous, with a separation of the 3 bases. Conical inflorescence apex or axillary. The flowers are small, the green white; 6 flowers and basal apart, the 2 sides are soft. There are 2 rounds of flower medicine introversion. It can give birth to 9 stamens and align in 3 rows. 2 rounds of the anthers are introverted, and 3 rounds of anthers are extroverted. The anthers are ovate, 4 chambers, and the petals split, and the filament is short. There is still 1 wheel degenerate stamens in the most. 1 pistil, ovary upper, 1 chamber, 1 ovule. Berry fruit, ovate. The florescence is from August to November, and the fruit period is from November to the following year in February.

Resources Situation: It grows in thin forests with sunshine. It is distributed in Guangdong, Guangxi, Jiangxi, Zhejiang, Fujian, etc.

Medicinal Part: Skin, root, leaf.

Functions: Root, warming spleen and stomach for dispelling cold, expelling rheumatism. It is used to cure poor appetite, abdominal distension, diarrhea, abdominal pain and rheumatism, sore and swell, traumatic injury. Leaf, it is used to cure rheumatism, bone pain, cold dampness, diarrhea and stomachache.

Main Ingredients: The bark contains volatile oil like cinnamaldehyde, eugenol, safrole, etc. The leaf contains volatile oil like eugenol, linalool.

8. 土茯苓

***Smilax glabra* Roxb.**

科属：百合科，菝葜属。

药名：土茯苓。

别名：禹余粮、白余粮、光叶菝葜。

植物特征：攀缘状灌木。根茎块根状，有明显结节，着生多数须根。茎无刺。单叶互生；革质，披针形至椭圆状披针形，先端渐尖，基部圆形，全缘，下面常被白粉，基出脉3~5条；叶柄长1~2 cm，略呈翅状，近基部具开展的叶鞘，叶鞘先端常变成2条卷须。花单性，雌雄异株；伞形花序腋生，花序梗极短；小花梗纤细，基部有多数宿存的三角形小苞片；花小，白色；花被裂片6，2轮；雄花的雄蕊6，花丝较花药短，退化雌蕊缺；雌花的退化雄蕊线形，子房上位，3室，柱头3歧，稍反曲。浆果球形，红色。花期7~11月，果期11月至次年4月。

资源状况：生长于山坡、荒山及林边的半阴地。分布于广东、湖南、湖北、浙江、四川、安徽、福建、江西、广西、江苏等地。

药用部位：根茎。夏、秋两季采挖，除去须根，洗净，干燥；或趁鲜切成薄片，干燥。

功能主治：除湿，解毒，通利关节。用于湿热淋浊、带下、痈肿、瘰疬、疥癣、梅毒及汞中毒所致的肢体拘挛、筋骨疼痛等。

主要成分：根茎含落新妇苷、黄杞苷、3-O-咖啡酰莽草酸、莽草酸、阿魏酸、β-谷甾醇、葡萄糖。

Genus：Liliaceae, Smilax.

Scientific Name：*Smilax glabra* Roxb..

English Name：Rhizoma smilacis glabrae.

Features：Climbing shrub. Root tuber root, with clear nodules, born mostly with fibril. Stems without a thorn. Single leaf alternate. Leathery, lanceolate to elliptical lanceolate, apex acuminate, base round, whole margin, below are often white, 3-5 basal veins. Petiole is 1-2 cm long, slightly finned, and the base has a scabbard, which often turns into 2 tendrils. The flower is monotropic, dioecious strain. Umbellate inflorescence axillary, short peduncle. Petiole slender, base with triangular bracts of majority remaining. Flowers are small, white. The flower perianth cesma is 6, 2 rounds. The stamens of male flowers are 6, the filament is shorter than anther, degenerate pistil is lacking. Degenerate stamen linear form of female flowers, upper part of ovary, 3 chambers, 3 column heads, slightly antiqu. Berries are spherical, red. The florescence is from July to November, and the fruit period is from November to the following year in February.

Resources Distribution：It grows in the hillside, barren mountain and graveyard beside the forest. It is distributed in Guangdong, Hunan, Hubei, Zhejiang, Sichuan, Anhui, Fujian, Jiangxi, Guangxi, Jiangsu, etc.

Medicinal Part：Rhizome. Excavation in summer and autumn, remove fibrous root, rinse and dry out, or take advantage of the fresh sliced, dry out.

Functions：Dispelling dampness and removing toxicity, easing tension of joint. It is used to cure hot, humid and drench turbidity, leucorrhea, carbuncle swollen, scrofulosis, acariasis, syphilis, limb spasm and bone pain caused by mercury poisoning.

Main Ingredients：The rhizome contains astilbe glucoside, astragalin, 3-O-Coffee acyl shikimic acid, shikimic acid, ferulic acid, β-sitosterin, glucose.

9. 射 干

***Belamcanda chinensis* (L.) Redouté.**

科属：鸢尾科，射干属。

药名：射干。

别名：乌扇、扁竹、交剪草、野萱花。

植物特征：多年生草木，根茎鲜黄色，须根多数。茎直立。叶2列，扁平，嵌迭状广剑形，绿色，常带白粉，先端渐尖，基部抱茎，叶脉平行。总状花序顶生，二叉分歧；花梗基部具膜质苞片，苞片卵形至卵状披针形，长1 cm左右；花被6，2轮，内轮3片较小，花被片椭圆形，先端钝圆，基部狭，橘黄色而具有暗红色斑点；雄蕊3，短于花被，花药外向；子房下位，3室。花柱棒状，柱头浅3裂。蒴果椭圆形，具3棱，成熟时3瓣裂。种子黑色，近球形。花期6~8月，果期7~9月。

资源状况：生长于山坡、草原、田野旷地，或为栽培。分布于全国各省。主产于湖北、河南、江苏、安徽、湖南、浙江、贵州、云南等地。

药用部位：根茎。

功能主治：清热解毒，消痰，利咽。用于热毒痰火郁结、咽喉肿痛、痰涎壅盛、咳嗽气喘等。

主要成分：根茎含射干定、鸢尾甙、鸢尾黄酮甙、鸢尾黄酮。花、叶含芒果甙。

Genus: Iridaceae, Belamcanda.

Scientific Name: *Belamcanda chinensis* (L.) Redouté..

English Name: Rhisoma belamcandae.

Features: Herbs are perennial. Rhizome is bright yellow, mostly is fibril. Stems are erect. Leaves are 2 columns, flat, imbedded broad sword shape, green, often white powder, apex acuminate, base hug stem, leaf veins parallel. The general inflorescence is terminal with bifurcation difference. Peduncle base with membranous bract, bracts ovate to lanceolate, oval 1 cm long. Perianth 6, 2 wheels, the inner wheel has 3 smaller pieces, tepals elliptic, apex obtuse, base narrow, orange and has a dark red spots; 3 stamens, shorter than perianth, anther extroversion; Subchamber, 3 Chambers. Flower column rod, stigma shallow has 3 cracks. Capsule elliptic, with 3 edges, 3 lobes when mature. The seeds are black and nearly spherical shape. The florescence is from June to August. The fruit period is from July to September.

Resources Situation: It grows in the hillside, prairie and open land in the field, or is cultivated. It is produced in all provinces of China. It is mainly produced in Hubei, Henan, Jiangsu, Anhui, Hunan, Zhejiang, Guizhou, Yunnan, etc.

Medicinal Part: Rhizome.

Functions: Clearing heat and removing toxicity, dissolving phlegm, relieving sore throat. It is used to cure toxic heat and phlegm, sore throat, excessive phlegm, cough and asthma.

Main Ingredients: The rhizome contains belamcandin, iridin, tectoridin, tectorigenin. The flower and leaf contain mangiferin.

10. 郁 金

***Curcuma aromatica* Salisb.**

科属：姜科，姜黄属。

药名：郁金。

别名：玉金、白丝郁金、毛姜黄、马莛、黄郁。

植物特征：多年生宿根草本。根粗壮，末端膨大成长卵形块根。块茎卵圆状，侧生，根茎圆柱状，断面黄色。叶基生，具叶耳；叶片长圆形，先端尾尖，基部圆形或三角形，叶背无毛。穗状花序；具鞘状叶，基部苞片阔卵圆形，小花数朵，生于苞片内，顶端苞片较狭，腋内无花；花萼白色筒状，不规则3齿裂；花冠管呈漏斗状，裂片3，纯白色而不染红，上面1枚较大，两侧裂片长圆形；侧生退化雄蕊长圆形，药隔距形，花丝扁阔；子房被伏毛，花柱丝状，基部有2棒状附属物，柱头略呈2唇形，具缘毛。花期4~5月。

资源状况：分布于江苏、浙江、福建、广东、广西、江西、四川、云南等地。

药用部位：干燥块根。

功能主治：行气化瘀，清心解郁，利胆退黄。用于治疗经闭痛经，胸腹胀痛、刺痛，热病神昏，癫痫发狂，黄疸尿赤等。

主要成分：郁金块根含挥发油，主要为姜黄烯、倍半萜烯醇等。其有效成分是对甲苯基-甲基羟甲基姜黄素。

Genus: Zingiberaceae, Curcuma.

Scientific Name: *Curcuma aromatica* Salisb..

English Name: Tuber curcumae.

Features: Perennial root herbs. Rough root, the bottom is expand to ovoid earthnut. Tuber oval, lateral, rhizome cylindric, the fracture surface is yellow. Leaf base, has leaf ear; Leaf blade is circular, the apex has sharp bottom, the base is circle or triangle. The back of blade is without hair. Spica inflorescence; It is sheath leaf shape, the base bracts are broad ovoid shape, and it has a number of flowers, born in bracts, the apex bracts are narrower, no flowers in axillary; Calyx white tube shape, irregular 3 teeth cracks; Corolla tube is funnel shaped, lobes 3, pure white without red, one on the top is bigger, the two sides crack is long circle shape. The lateral degenerative stamens are round, the drug isolation is rectangle shape, and the filament is flat and wide. Ovary is fleeced, flower-shaped, the base has 2 lolliform appendages, the stigma is slightly like 2-lipped shape, and it has edge wool. The florescence is from April to May.

Resources Situation: It is distributed in Jiangsu, Zhejiang, Fujian, Guangdong, Guangxi, Jiangxi, Sichuan, Yunnan, etc.

Medicinal Part: Dried root.

Functions: Circulating Qi and relieving stasis, calming heart and expelling depression, normalizing the gallbladder function and treat jaundice. It is used to cure amenorrhea dysmenorrhea, chest pain, stabbing pain, fever and dizzy, epilepsy crazy, jaundice and urinary pain.

Main Ingredients: The root contains volatile oil, mainly is the curcumene, sesquiterpene alcohol, etc. The effective component is cresyl-methyl hydroxy methyl curcumin.

11. 莪术

***Curcuma zedoaria* (Christm.) Rosc.**

科属：姜科，姜黄属。

药名：莪术。

别名：山姜黄、蓬莪茂。

植物特征：多年生草本，全株光滑无毛。叶椭圆状长圆形至长圆状披针形，中部常有紫斑；叶柄较叶片为长。花茎由根茎单独发出，常先叶而生；穗状花序；苞片多数，下部的绿色，缨部的紫色；花萼白色，顶端3裂；花冠黄色，裂片3，不等大；侧生退化雄蕊小；唇瓣黄色，顶端微缺；药隔基部具叉开的矩。蒴果三角形。花期4~6月。

资源状况：生长于山谷、溪旁及林边等的阴湿处。主产于广西、四川。

药用部位：根茎。根茎称"莪术"，块根称"绿丝郁金"。

功能主治：行气破血，消积止痛。用于癥瘕痞块、瘀血经闭、食积胀痛、早期宫颈癌等治疗。

主要成分：根茎含挥发油（莪术呋喃酮、表莪术呋喃酮、莪术呋喃烃、莪术双酮、莪术醇、樟脑、龙脑等）。

Genus：Zingiberaceae, Curcuma.

Scientific Name：*Curcuma zedoaria* (Christm.) Rosc..

English Name：Khizoma curcumae.

Features：Perennial herbs. The whole stool is smooth and glabrous. Leaf is elliptic long round to long circular lanceolate shape, usually purple in the middle. Petioles are longer than leaves. Cauliflower is emitted by roots alone, often leaves born first; Spica inflorescence; A lot of bracts, the lower part is green, and the tassel part is purple; White calyx, the apex has 3 cracks; Yellow corolla, 3 lobes, different size; Lateral degeneration stamens are small; Yellow lip petals, the apex is scare slightly; The base has a divergent rectangle shape. The capsule is triangle shape.The florescence is from April to June.

Resources Situation：It grows in valley, wet places along the stream side and forests. It is mainly produced in Guangxi, Sichuan.

Medicinal Part：Rhizome. The rhizome is also called Curcuma zedoania (Christm.) rosc. The earthnut is also called "Green silk turmeric".

Functions：Circulating Qi and blood, removing food retention and relieving pain. It is used to cure lump and blood stasis, amenorrhea, dyspepsia and swelling pain, early stage cervical carcinoma.

Main Ingredients：The root contains volatile oil such as curzerenone, furan ketone, rhizoma zedoariae furan hydrocarbon, rhizoma zedoariae diketone, curcumenol, camphor, borneol, etc.

12. 余甘子

***Phyllanthus emblica* Linn.**

科属：大戟科，叶下珠属。

药名：庵摩勒。

别名：余甘子、滇橄榄、油柑子。

植物特征：乔木。树皮灰白色，薄而易脱落，露出大块赤红色内皮。叶互生于细弱的小枝上，2列，密生，极似羽状复叶；近无柄；落叶时整个小枝脱落；托叶线状披针形；叶片长方线形或线状长圆形。花簇生于叶腋，花小，黄色；单性，雌雄同株，具短柄；每花簇有1朵雌花，花萼5~6片，无瓣；雄花花盘成6个极小的腺体，雄蕊3，合生成柱；雌花花盘杯状，边缘撕裂状，子房半藏其中。果实肉质，径圆而略带6棱，初为黄绿色，成熟后呈赤红色，味先酸涩而后回甜。花期4~6月，果期7~9月。

资源状况：生长于海拔300~1 200 m的疏林下或山坡向阳处。分布于福建、台湾、广东、海南、广西、四川、贵州、云南等地。

药用部位：成熟果实。

功能主治：清热凉血，消食健胃，生津止咳。用于治疗血热血瘀、消化不良、腹胀、咳嗽、喉痛、口干等。

主要成分：果实含鞣质，干果含粘酸。果皮含没食子酸、油柑酸、余甘子酚。种子含亚麻酸、亚油酸、油酸等。

Genus: Euphorbiaceae, Phyllanthus.

Scientific Name: *Phyllanthus emblica* Linn..

English Name: Yunnan olive, myrobalan.

Features: Dungarunga. Bark is gray, thin and easy to fall off, revealing a large red inner skin. Leaves alternate between small, thin branches, 2 columns, dense, and extremely featherlike compound leaves; Nearly sessile; The whole branchlets fall off when falling leaves; Torus linear lanceolate; Leaf is blade long square linear or linear long circle. Flower clusters in axillary, flowers small, yellow; The flower is monotropic, monoecism, with short stalks; Each flower cluster has 1 female flower, calyx 5-6 tablets, no flap; The male flower is formed into 6 small glands, 3 stamens, and the composite is formed. Female flower is cup shape, edge tear, ovary half hidden. The fruit fleshy, the diameter circle and slightly 6 edges, the first yellow green, then turn ripe red when mature, taste sour and sweet. The florescence is from April to June. The fruit period is from July to September.

Resources Situation: It grows in the thin forests or hillsides with sunshine of 300-1, 200 m. It is distributed in Fujian, Taiwan, Guangdong, Hainan, Guangxi, Sichuan, Guizhou, Yunnan, etc.

Medicinal Part: Mature fruit.

Functions: Removing heat and cooling blood, improving digestion, helping produce saliva and relieving cough. It is used to cure blood heat and stasis, dyspepsia, abdominal distension, cough, sore throat, thirst.

Main Ingredients: The fruit contains tannin, and dried fruit contains mucic acid. The pericarp contains gallic acid, phyllemblic acid, emblicol. The seed contains linolenic acid, linoleic acid, oleic acid, etc.

13. 断肠草

Gelsemium elegans (Gardn. et Champ.) Benth.

科属：马钱科，钩吻属。

药名：断肠草。

别名：秦钩吻、胡蔓藤、大茶药、吻莽。

植物特征：常绿木质藤本，枝光滑。叶对生，卵状长圆形至卵状披针形，先端渐尖，基部楔形或近圆形，全缘；叶柄长1.2 cm。3歧分枝的聚伞花序，顶生或腋生；花小，黄色；苞片2，小而狭；萼片5，分离；花冠漏斗状，先端5裂，内有较淡的红色斑点，裂片卵形，先端尖，较花筒为短；雄蕊5；子房上位，2室，花柱丝状，柱头4裂。蒴果卵状椭圆形，分裂为2个2裂的果瓣。种子多数有翅。花期5~11月，果期7月至次年2月。

资源状况：生长于向阳的山坡、路边的草丛或灌丛中。分布于浙江、福建、广东、广西、贵州、云南等地。

药用部位：全草。

功能主治：祛风，攻毒，消肿，止痛。治疥癞、湿疹、瘰疬、痈肿、疔疮、跌打损伤、风湿痹痛、神经痛等症。

主要成分：根、茎、叶含钩吻碱子、寅、卯、甲、丙、辰、戊、丁等8种生物碱。其中钩吻碱子的含量最高，钩吻碱剧毒，为最重要的有效成分。

Genus: Loganiaceae, Gelsemium.

Scientific Name: *Gelsemium elegans* (Gardn. et Champ.) Benth..

Features: Evergreen wooden rattan, smooth branches. Leaves opposite, ovate-long and ovate-lanceolate shape, the apex is acuminate, base cuneate or nearly round, whole margin. The petiole is 1.2 cm long. The cytoplasa of the branches, basifixed or axillary; Small flowers, yellow; 2 bracts, small and narrow; 5 sepals, dissociation; Corolla funnel, the apex has 5 clefts, with red spots inside, lobes ovate shape, the apex is sharpe, shorter than the corolla; 5 stamens; Upper ovary, 2 chambers, the stylus is filiform shape, 4 stigmas. Capsule is ovoid shape, divided into 2 cracks fruit petals. Most seeds have wings. The florescence is from May to November. The fruit period is from July to the following year in February.

Resources Situation: It grows in the apricus hillside, tussock or bush along the road. It is distributed in Zhejiang, Fujian, Guangdong, Guangxi, Guizhou, Yunnan, etc.

Medicinal Part: The whole herb.

Functions: Expelling wind, curing poison, detumescence and relieving pain. It is used to cure scabies infection, eczema, scrofulosis, welling-abscess swelling, furunculosis, traumatic injury, rheumatism, arthralgia spasm pain, neuralgia.

Main Ingredients: The root, rhizome, leaf contains 8 alkaloids hook elements Zi, Yin, Mao, Jia, Bing, Chen, Wu, Ding. The koumine contains the most and the alkaloids hook Yan is poisonous and as the most effective ingredient.

14. 土牛膝

Achyranthes aspera Linn.

科属：苋科，牛膝属。

药名：土牛膝。

别名：杜牛膝、倒扣草、倒钩草。

植物特征：多年生草本。根细长，外皮土黄色。茎直立，四棱形，贝条纹，疏被柔毛，节略膨大，节上对生分枝。叶对生，叶柄长5~20 mm；叶片椭圆形或椭圆状披针形，先端长尖，基部楔形或广楔形，全缘，两面被柔毛。穗状花序腋生兼顶生，初时花序短，花紧密，其后伸长；花皆下折贴近花梗；苞片1，膜质，宽卵形，上部突尖成粗刺状，另有2枚小苞片针状，先端略向外曲，基部两侧各具，1卵状膜质小裂片；花被绿色，5片，直立，披针形，有光泽，具1脉，边缘膜质；雄蕊5，花丝细，基部合生，花药卵形，2室，退化雄蕊顶端平或呈波状缺刻；子房长圆形，花柱线状，柱头头状。胞果长圆形，光滑。种子1枚，黄褐色，卵形。花期6~8月，果期10月。

资源状况：栽培或野生于山野路旁。分布于河南、山西、山东、江苏、安徽、浙江、江西、湖南、湖北、四川、云南、贵州等地。主产于河南。

药用部位：根和根茎。

功能主治：活血散瘀，祛湿利尿，清热解毒。治淋病、尿血、妇女经闭、症瘕、风湿关节痛、脚气、水肿、痢疾、疟疾、白喉、痈肿、跌打损伤等。

主要成分：根含皂甙、脱皮甾酮和牛膝甾酮。

Genus: Amaranthaceae, Achyranthes.

Scientific Name: *Achyranthes aspera* Linn..

Features: Perennial herbs. The roots are long and thin, and the skin is yellow. Stems erect, quadrigonal, striate, sparse vegetation is the pubescence, the section is enlarge, joint bifurcated. Leaf opposite, petiole is about 5-20 mm; Leaf blade elliptic or elliptical lanceolate, the apex is long and spare, the base is cuneiform or broad wedge, full margin, both sides are pubescence. Flower heads in the axillary and apex, with short inflorescence, close and elongated. The flowers are folded down to the peduncle; 1 bract, membranous, broad-ovate, supersharp, with 2 small bracts spicules, apex slightly outward, base on both sides, 1 ovoid membranous lobes; Flowers are green, 5 pieces, erect, lanceolate, glossy, 1 vein, margin membranous; 5 stamens, filaments are thin, base congenies, anthers ovate, 2 chambers, degenerate stamens apical apical or wavy; Long and round ovary, cylindrical linear. Stigma is head shape. Cytoplasm is round and smooth. 1 seed and yellow brown, ovoid. The florescence is from June to August, and the fruit period is in October.

Resources Situation: It is cultivated or wildly grows in the roadside of mountains and plains. It is distributed in Henan, Shanxi, Shandong, Jiangsu, Anhui, Zhejiang, Jiangxi, Hunan, Hubei, Sichuan, Yunnan, Guizhou, etc. It is mainly produced in Henan.

Medicinal Part: Root and rhizome.

Functions: Circulating blood and eliminating stasis, clearing dampness, diuresis, clearing heat and removing toxicity. It is used to cure gonorrhoea, hematuria, amenorrhea, abdominal mass, rheumatoid joint pain, dermatophytosis, edema, dysentery, malaria, diphtheria, welling-abscess swelling, traumatic injury.

Main Ingredients: The root contains saponin, ecdysterone and inokosterone.

15. 穿破石

Maclura tricuspidata Carrière

科属：桑科，柘属。

药名：穿破石。

别名：柘根、川破石。

植物特征：落叶灌木或小乔木。根皮柔软，黄色。枝灰褐色，光滑，皮孔散生，具粗壮、直立或微弯的棘刺。单叶互生，革质，倒卵状披针形、椭圆形或长椭圆形，先端钝或渐尖，基部楔形，全缘，两面无毛。花单性，雌雄异株；头状花序单生或成对，具短柄，被柔毛；雄花序花被片3~5，楔形，不相等，被毛；雌花序球状，结果时增大，花被片4，顶端厚，被茸毛。聚花果球形，肉质，直径达2.5 cm，被毛，瘦果包裹在肉质的花被和苞片中，成熟时橘红色。花期5~6月，果期6~7月。

资源状况：生长于山坡、溪边、灌丛中。分布于湖南、安徽、浙江、福建、广东、广西等地。主产于广东、广西、福建等地。

药用部位：根。

功能主治：祛风利湿，活血通经。治风湿关节疼痛、黄疸、淋浊、蛊胀、闭经、劳伤咯血、跌打损伤、疔疮痈肿等症。

主要成分：含黄酮甙、酚类、氨基酸、有机酸、糖类。

Genus：Moraceae, Madura.

Scientific Name：*Maclura tricuspidata* Carrière.

Features：Deciduous shrub or dungarunga. The root skin is soft and yellow. Branches are grayish brown, smooth, pinhole scattered, with a stout, erect or slightly curved ratchet. Single leaf alternate, leathery, ovate lanceolate, elliptic or long elliptic, the apex is obtuse or acuminate, base cuneiform, full margin, both sides glabrous. The flowers are monotropic, dioecism. The capitulum is single or pair, with short stalks, pubescence. Male inflorescence are 3-5 pieces, cuneiform, different sizes, hairy. Female flower is inflorescence, it will enlarge when bear fruit, the flower has 4 pieces, the top is thick, has fuzz. Globular, fleshy, 2.5 cm in diameter, covered with fleshy floral perianth and bracts. The florescence is from May to June. The fruit period is from June to July. It is nacarat when it is ripe.

Resources Situation：It grows on the hillside, by the stream and bush. It is distributed in Hunan, Anhui, Zhejiang, Fujian, Guangdong, Guangxi and is mainly produced in Guangdong, Guangxi, Fujian, etc.

Medicinal Part：Root.

Functions：Dispelling wind and dampness, circulating blood and promoting collaterals. It is used to cure rheumatoid joint pain, jaundice, stranguria with turbid discharge, tympanites due to parasitosis, amenorrhea, internal lesion caused by overexertion coughing blood, traumatic injury, furuncle carbuncle swollen.

Main Ingredients：Flavonoid glycosides, phenols, amino acid, organic acid, carbohydrate.

16. 九里香

***Murraya exotica* L.**

科属：芸香科，九里香属。

药名：九里香。

别名：千里香、满山香、石桂树。

植物特征：灌木或乔木，木材极硬。单数羽状复叶；小叶互生，3~7枚，有时退化为1枚；小叶变异大，由卵形、匙状倒卵形、椭圆形至近菱形，先端钝或钝渐尖，有时稍稍凹入，基部阔楔尖或楔尖，有时略偏斜，全缘。伞房花序短，顶生或生于上部叶腋内，通常有花数朵，花白色，极芳香；花萼极小，5深裂；花瓣5，分离，覆瓦状排列；雄蕊10，花丝柔弱；子房上位，2室，花柱柔弱，柱头头状。果卵形或球形，肉质，红色，先端尖锐，有种子1~2颗。花期4~8月，果期9~12月。

资源状况：生长于山野中，亦有栽培者。产于广东、广西、福建等南部地区。

药用部位：干燥叶、带叶嫩枝及根。

功能主治：行气止痛，活血散瘀。用于治疗胃痛、风湿痹痛等症；外治牙痛、跌扑肿痛、虫蛇咬伤等。

主要成分：含多种香豆精类化合物（如九里香甲素、九里香乙素、九里香丙素、长叶九里香内酯二醇、长叶九里香醛等）。

Genus: Rutaceae, Murraya.

Scientific Name: *Murraya exotica* L..

English Name: Folium et cacumen murrayae.

Features: Shrubs or trees, hard wood. Simple pinnate compound leaf; Leaves alternate, 3-7 pieces, sometimes degenerate to 1 piece; Lobular mutation is large, from ovate, spoonful, oval-shaped, elliptic to rhomboid, the apex is obtuse or blunt acuminate, sometimes slightly concave, base broad wedge tip or wedge tip, sometimes slightly oblique, whole edge. The inflorescence of the umbrella is short, the head is born in the upper leaf axillary, usually have several flowers, white, extremely aromatic; Calyx is small, 5 deep fissures; 5 petals, detached, covered in tiles; 10 stamens, filaments weak; Ovary upper, 2 chambers, flower column soft, stigma head. Fruit ovate or spherical, fleshy, red, apex acute, with 1-2 seeds.The florescence is from April to August. The fruit period is from September to December.

Resources Situation: It grows or is cultivated in the mountains. It is produced in the south of China like Guangdong, Guangxi, Fujian, etc.

Medicinal Part: Dried leaf, twig with leaf and root.

Functions: Circulating Qi and relieving pain, circulating blood and relieving stasis. It is used to cure stomachache, reheumatism, arthralgia spasm pain and external use for toothache, traumatic injury, snake bite.

Main Ingredients: Many coumarin compounds such as isomexoticin, murpanidin Yi, murpanidin Bing, leaf lactone diol murraya ointment, leaf lactone diol murraya aldehyde, etc.

17. 三叉苦

***Evodia lepta* (Spreng.) Merr.**

科属：芸香科，吴茱萸属。

药名：三叉苦。

别名：三叉虎、三丫苦、三桠苦。

植物特征：落叶灌木或小乔木。树皮灰白色，不剥落；嫩芽被短毛，余秃净。叶对生；指状复叶；小叶3片，矩圆形或长椭圆形，纸质，先端长尖，基部渐狭而成一短柄，全缘。花单性，圆锥花序，腋生，有近对生而扩展的分枝，被短柔毛；小苞片三角形；花萼4，矩圆形，被短毛；花瓣4，黄色，卵圆形；雄花的雄蕊4枚，长过花瓣1倍；雌花的子房上位，4室，被毛，花柱有短毛，柱头4浅裂。果由4个分离的心皮所构成，中间有发育不健全的1～3个心皮。种子黑色，圆形，有光泽。花期5～6月，果期6～8月。

资源状况：生长于山谷、溪边、林下。分布于我国南部各地。

药用部位：叶、根或根皮。

功能主治：清热，解毒，祛风，除湿。治咽喉肿痛、疟疾、黄疸型肝炎、风湿骨痛、湿疹、皮炎、疮疡等症。

主要成分：含生物碱。

Genus: Rutaceae, Evodia.

Scientific Name: *Evodia lepta* (Spreng.) Merr..

Features: Deciduous shrub or dungarunga. The bark is gray white, not peeling. The tender buds are lover with short and bare. Leaves opposite; Finger compound leaf; Leaflet are 3 pieces, rectangular or long oval, papery, apex long pointed, base gradually narrow and become a short stalk, whole edge. Flower unisex, panicle, axillary, with a branch of the near opposite to the extension, pubescence; Bracts triangle; 4 calyxes, rectangle and round shape, cover with short hair; 4 petals, yellow, ovoid; 4 male stamens, 1 time longer than petals.The ovary of the female flower is upper, 4 chambers, is hairy, the style of the flower has short hair, the stigma has 4 supersulcus. The fruit consists of four separate carpeted hearts, with an underdeveloped 1-3 carpels. Seeds are black, round, glossy. The florescence is from May to June. The fruit period is from June to August.

Resources Situation: It grows in the valley, forest and by the stream. It is distributed in Southern China.

Medicinal Part: Leaf, root or velamen.

Functions: Clearing heat and removing toxicity, expelling wind and dampness. It is used to cure sore throat, malaria, jaundice hepatitis, rheumatism and bone pain, eczema, dermatitis, sore and ulcer.

Main Ingredients: Alkaloid.

18. 紫金牛

***Ardisia japonica* (Thunb.) Blume**

科属：紫金牛科，紫金牛属。

药名：紫金牛。

别名：平地木、叶下红。

植物特征：常绿小灌木。地下茎作匍匐状，具有纤细的不定根。茎单一，圆柱形，表面紫褐色，有细条纹，具有短腺毛。叶互生，通常3～4叶集生于茎梢，呈轮生状；叶柄长5～10 mm，密被短腺毛；无托叶；叶片椭圆形，先端短尖，边缘具细锯齿，基部楔形，上面绿色，有光泽，下面淡紫色，老时带革质。花着生于茎梢或顶端叶腋，2～6朵集成伞形；花两性；花萼5裂，裂片三角形；花冠白色或淡红色，5深裂，裂片卵形而先端锐尖，两面无毛，具有赤色斑点；雄蕊5，雌蕊1，子房球形，花柱细，顶端尖而弯曲。核果，球形，熟时红色，经久不落。花期5~6月，果期11~12月。

资源状况：分布于福建、江西、湖南、四川、江苏、浙江、贵州、广西、云南等地。

药用部位：茎叶。

功能主治：镇咳，祛痰，活血，利尿，解毒。治慢性气管炎、肺结核咳嗽咯血、吐血、脱力劳伤、筋骨酸痛、肝炎、痢疾、急慢性肾炎、高血压、疝气、肿毒等症。

主要成分：全株含挥发油、岩白菜素。还含有2-羟基-5-甲氧基-3-十五烯基苯醌等化合物及三萜类化合物。叶中含有槲皮甙、杨梅树皮甙和冬青萜醇。

Genus: Myrsinaceae, Ardisia.

Scientific Name: *Ardisia japonica* (Thunb.) Blume.

Features: Evergreen shrubs. The rhizome is prostrate, with slender and untenable roots. Stem is single, cylindrical, surface purple, with fine stripes, with short glandular hairs. Leaves alternate, usually 3-4 leaves are born in the stem tip, in rotation; Petiole is 5-10 mm long, densely covered with short glandular hairs; Without stipule; Leaf blade elliptic, the apex is short and sharp, margin serrate, base cuneate, upper green, glossy, violet below, leathery at old age. Flowers are born from stem tip or apical axillary, 2-6 pieces integrate into umbrella form; Flowers bisexual; The calyx has 5 clefts, lobes triangular; Corolla white or reddish, 5 deep cleft, lobes ovate and apex acute, glabrous, with red spots; 5 stamens; 1 pistil. Ovary spherical, spiky, the apex is sharp and bent. Drupe, globular, red when it is mature, and durable. The florescence is from May to June, and the fruit period is from November to December.

Resources Situation: It is distributed in Fujian, Jiangxi, Hunan, Sichuan, Jiangsu, Zhejiang, Guizhou, Guangxi, Yunnan, etc.

Medicinal Part: Steam leaf.

Functions: Relieving cough, eliminating phlegm, circulating blood, diuresis, removing toxicity. It is used to cure chronic bronchitis, tuberculosis cough haemoptysis, vomit blood, internal lesion caused by overexertion, aching muscles, hepatitis, dysentery, acute or chronic nephritis, hypertension, hernia, pyogenic infections.

Main Ingredients: Volatile oil, chemical compound and triterpenes compound like bergenin, 2-hydroxyl -5-methoxy-3-fifteen alkenyl benzoquinone. The leaf contains quercitrin, myricitrin, bergenin and holly terpene alcohols.

19. 白花蛇舌草

***Hedyotis diffusa* Willd.**

科属：茜草科，耳草属。

药名：白花蛇舌草。

别名：蛇舌草、蛇舌癀、二叶葎。

植物特征：一年生草木。茎纤弱，略带方形或圆柱形，无毛。叶对生，叶片线形至线状披针形，革质；托叶膜质，基部合生成鞘状，顶端有细齿。花单生或2朵生于叶腋；花萼筒状，4裂，裂片边缘被短刺毛；花冠漏斗形，纯白色，先端4深裂；雄蕊4；子房2室，柱头2浅裂呈半球状。蒴果，扁球形，室背开裂，花萼宿存。种子棕黄色，极细小。花期春季。

资源状况：生长于山坡、路边、溪畔草丛中。分布于云南、广东、广西、福建、浙江、江苏、安徽等地。

药用部位：全草。

功能主治：清热解毒，利尿消肿，活血止痛。用于肠痈（阑尾炎）、疮疖肿毒、湿热黄疸、小便不利等症；外用治疮疖痈肿、毒蛇咬伤。

主要成分：全草中分出卅一烷、豆甾醇、熊果酸、齐墩果酸、β-谷甾醇、β-谷甾醇-D-葡萄糖甙、对香豆酸等。

Genus: Rubiaceae, Hedyotis.

Scientific Name: *Hedyotis diffusa* Willd..

English Name: Spreading hedyotis herb.

Features: Stems slender, slightly square or cylindrical, glabrous. Leaf opposite, leaf blade linear to linear lanceolate, leathery. The topical membrane, the base of a sheath, the apex has teeth. Flower is single-birth or has 2 pieces born in the axillary; 4 calyx tubes, 4 clefts, margin of lobes is covered with short puncture; Corolla funnel shaped, pure white, the apex has 4 deep cracks; 4 stamens; 2 ovaries. The stigma has two cracks and is semi-spherical. Capsule, flat globular, ventricle cracked, calyx persistent. The seeds are brown and small. The florescence is in spring.

Resources Situation: It grows in the hillside, roadside, tussock of the stream. It is distributed in Yunnan, Guangdong, Guangxi, Fujian, Zhejiang, Jiangsu, Anhui, etc.

Medicinal Part: The whole herb.

Functions: Clearing heat and removing toxicity, inducing diuresis and edema, circulating blood and relieving pain. It is used to cure appendicitis, sore furuncle and swollen poison, damp heat and jaundice, difficult urination and external use for sore furuncle carbuncle swollen and snake bite.

Main Ingredients: The whole grass contains alkanes, stigmasterol, ursolic acid, oleanolic acid, β-sitosterol, β-sitosterol-D-glucoside, p-coumaric acid, etc.

20. 广防己

***Aristolochia fangchi* Y.C.Wu ex L.D.Chow et S.M.Hwang**

科属：马兜铃科，马兜铃属。

药名：广防己。

别名：木防己、藤防己。

植物特征：多年生木质藤本，长3~4 m；根粗壮，圆柱形，栓皮发达，长15 cm以上；茎下部不分枝，树皮厚，纵裂松软，嫩枝密被褐色长柔毛；叶互生，革质长圆形或卵状长圆形，主脉3条，基出，下面网脉凸起；总状花序，紫色，喇叭形，花被合生，紫红色，有黄斑及网纹，外被褐色绒毛，花被管基部膨大，上部短小呈檐部圆盘状，边缘3浅裂；蒴果圆柱形或长圆状披针形；种子褐色，多数。花期3~5月，果期7~9月。

资源状况：广防己主产于广东、广西和云南等地。

药用部位：根。

功能主治：用于湿热身痛、下肢水肿、小便不利、风湿痹痛。

主要成分：含马兜铃酸、马兜铃酸B、马兜铃酸C、木兰碱等。

Genus: Aristolochiaceae, Aristolochia.

Scientific Name: *Aristolochia fangchi* Y.C. Wu ex L.D. Chow et S.M. Hwang.

English Name: Wood stephania, Rattan.

Features: Perennial wooden cane with 3-4 m long; Root stout, cylindrical, developed cork and longer than 15 cm; Stem not branched, thick bark, soft slime, tender branches densely brownish long pubes; alternate leaves, leathery long round or ovate oblong, 3 main veins, out of base, the following mesh veins bulge; General inflorescence, purple, horn shape, the flower is syngenetic, violet red, has the yellow spot and the net grain, the outside is the brown fluff, the flower is tubule of the base bulges, the upper part is small in eaves and disk, the edge with 3 shallow cracks. Fruit is cylindrical or long round lanceolate. Seeds are mostly brown. The florescence is from March to May, and the fruit period is from July to September.

Resources Situation: *Aristolochia fangchi* Y.C. Wu ex L.D. Chow et S.M. Hwang is mostly produced in Guangdong, Guangxi, Yunnan, etc.

Medicinal Part: Root.

Functions: Relieving hot and humid body pain, lower limb edema, dysuria and rheumatism pain.

Main Ingredients: Aristolochic acid, aristolochic acid B, aristolochic acid C, magnoline, etc.

21. 防风草

Epimeredi indica (L.) Rothm.

科属：唇形科，广防风属。

药名：防风草。

别名：落马衣、土防风、马衣叶。

植物特征：直立粗壮草本。茎方柱形，密被白色柔毛。叶对生，纸质，阔卵形或卵形，花冠淡紫色，在茎、枝上部排成多花的轮伞花序，通常再复合成顶生穗状花序，花冠向上渐变阔大，冠檐二唇形。小坚果圆球形，黑色。花期8~9月，果期9~11月。

资源状况：防风草主产于广东、广西、贵州、云南、西藏东南部、四川、湖南南部、江西南部、浙江南部、福建及台湾等地；生长于热带及南亚热带地区的林缘或路旁等荒地上。印度、东南亚经马来西亚至菲律宾也有。模式标本采自印度。

药用部位：全草。

功能主治：祛风解表，理气止痛。用于治疗感冒发热、风湿关节痛、胃痛、胃肠炎等症；外用于治疗皮肤湿疹、神经性皮炎、虫蛇咬伤、痈疮肿毒等症。

主要成分：含防风草内脂、挥发油及生物碱等。

Genus: Labiatae, Epimeredi.

Scientific Name: *Epimeredi indica* (L.) Rothm..

English Name: Parsnip.

Features: Erect with strong herbs. Stem square, densely white pubescent. Leaves opposite, papery, broadly ovate or ovate, pale purple corolla, in the upper stem and branches with more rounds of umbrella inflorescence, usually are composite to spica, corolla upward gradient, a large crown canopies with 2 lips. Small nut spheroidal shape, black. The florescence is from August to September. The fruit period is from September to November.

Resources Situation: *Epimeredi indica* (L.) Rothm. is mostly produced in Guangdong, Guangxi, Guizhou, Yunnan, southeast Tibet, Sichuan, south Hunan, south Jiangxi, south Zhejiang, Fujian, Taiwan, etc. It grows in wasteland of the tropical and subtropical forests or roadside. There are some in India, southeast Asia countries from Malaysia to Philippines. Type specimens are collected from India.

Medicinal Part: The whole herb.

Functions: Dispelling wind and presenting the circumstance, circulating Qi and preventing pain, relieving cold and fever, rheumatic arthralgia, stomach pain, gastroenteritis. For external use to treat skin eczema and neurodermatitis, snake bites, carbuncle sore and swollen poison.

Main Ingredients: Parsnip axunge, Essential oil and alkaloid, etc.

22. 八角枫

Alangium chinense **(Lour.) Harms**

科属：八角枫科，八角枫属。

药名：八角枫。

别名：八角金盘、木八角。

植物特征：落叶小乔木或灌木。树皮平滑，灰褐色。单叶互生，形状不一，常卵形至圆形，先端长尖，全缘或有2～3裂，裂片大小不一，基部偏斜，幼时两面均有毛，后仅脉腋处有丛毛和沿叶脉有短柔毛；主脉4～6条。聚伞花序腋生，具小花8～30余朵；苞片1，线形；萼钟状，有纤毛，萼齿6～8；花瓣与萼齿同数，白色，线形，反卷；雄蕊6～8；雌蕊1，子房下位，2室，花柱细圆筒形，有稀细毛，柱头3裂。核果黑色，卵形。花期5~7月和9~10月，果期7~11月。

资源状况：生长于山野或林中。分布于长江流域及南方各地。

药用部位：根。

功能主治：祛风，通络，散瘀，镇痛，并有麻醉及松弛肌肉作用。用于治疗风湿疼痛、麻木瘫痪、心力衰竭、劳伤腰痛、跌打损伤等症。

主要成分：含生物碱、酚类、氨基酸、有机酸、树脂。须根主要含生物碱及糖甙，又含强心甙。

Genus: Alangiaceae, Alangium.

Scientific Name: *Alangium chinense* (Lour.) Harms.

English Name: Fatsia japonica, Wood octagon.

Features: Small deciduous trees or shrubs. The bark is smooth and brown. Simple and alternate leaves, have different shapes, often ovate to rounded, apex long, margin entire or with 2-3 pieces, different lobes sizes, base oblique, as both sides are hairy, the axillary area has tuft and along the leaf veins have pubes. 4-6 main veins. A cluster of umbrellas with 8-30 flowers. 1 bract, linear; Calyx bell shape, cilia, 6-8 calyxes; The petals are the same number as the calyx, white, linear, and anti-roll. 6-8 stamens. 1 pistil, inferior ovary, 2 chambers, cylindrical thin cylindrical, with thin hairs, 3 stigmas. Nuclear fruit black, ovate. The florescence is from May to July, September to October. The fruit period is from July to November.

Resources Situation: It grows in mountains or plains, forests. It is distributed in the Yangtze river basin and the south areas of China.

Medicinal Part: Root.

Functions: Dispelling the wind, smoothing collaterals, dissipating blood stasis, analgesia, anesthesia and relaxing the muscles. It is used to cure pain from rheumatism, numbness paralysis, cardiac failure, internal lesion caused by overexertion lumbago and traumatic injury.

Main Ingredients: Alkaloids, phenols, amino acid, organic acid, resin. Fibrous root mainly contains alkaloids and glucoside, and also includes cardiac glycoside.

参考文献

[1] 中国植物志编委会. 中国植物志：1~80卷［M］. 北京：科学出版社，1961—2004.

[2] 中国科学院华南植物园. 广东植物志：1~7卷［M］. 广州：广东科技出版社，1987—2006.

[3] 国家药典委员会. 中华人民共和国药典：一部［M］. 北京：化学工业出版社，2005.

[4] 《全国中草药汇编》编写组. 全国中草药汇编：上下册［M］. 北京：人民卫生出版社，1982.

[5] 陈锡桥，吴七根. 澳门常见中草药：第一册［M］. 广州：广东科技出版社，2007.

[6] 王玉生，蔡岳文. 南方药用植物［M］. 广州：南方日报出版社，2011.

[7] 吴忠发，余丽莹. 广西中草药的资源分布情况及种类［J］. 重庆中草药研究，1999（2）：25-27.